STEPHEN H. KELLERT

in the wake of chaos

Unpredictable Order in Dynamical Systems

The University of Chicago Press
Chicago and London

Stephen H. Kellert is an assistant professor in the Department of the History and Philosophy of Science at Indiana Unviersity.

The University of Chicago Press, Chicago 60637
The University of Chicago Press, Ltd., London

Printed in the United States of America

02 01 00 99 98 97 96 95 94 93 12345
ISBN: 0-226-42974-1 (cloth)
0-226-42976-8 (paper)

Library of Congress Cataloging-in-Publication Data

Kellert, Stephen H.
In the wake of chaos : unpredictable order in dynamical systems / Stephen H. Kellert.
p. cm. — (Science and its conceptual foundations)
Includes bibliographical references and index.
1. Chaotic behavior in systems. 2. Science—Philosophy.
I. Title. II. Series.
Q172.5.C45K45 1993
003′.7—dc20 92-30355

∞ The paper used in this publication meets the minimum requirements of the American National Standard for Information Sciences—Permanence of Paper for Printed Library Materials, ANSI Z39.48-1984.

Contents

Acknowledgments

My first debt is to my teachers, beginning with Catherine Konefal and ending with those who guided me through graduate work: Arthur Fine, David Hull, and Stephen Toulmin. Special thanks to Arthur for his help and, yes, inspiration. Thanks also to Edie Fisher, Roberta Harvey, and Robert Alge, and to Georgia Warnke, R. I. G. Hughes, and John Franks. Thanks to Rick Jensen for introducing me to the study of dynamical chaos in his class at Yale in 1984.

Several people read various chapters at various stages and offered helpful suggestions, including Peter Kosso, Roger Jones, Mark Stone, Richard Sorrenson, Eric Winsberg, Jeff Kosokoff, and the members of my seminar on philosophy and chaos theory in the Spring of 1991. Opportunities to give presentations of some of the ideas in this book have also provided occasions for useful criticisms and advice from Linda Wessels, Noretta Koertge, John Winnie, and many others. I have benefited greatly from the careful readings and detailed comments of Jerry Gollub and an anonymous reviewer. I would like to thank Susan Abrams at the University of Chicago Press for all her help in making this book possible. Crucial encouragement along the way has come from Heinrich Von Staden, Meredith Williams, Tom Gieryn, and the Science and Literature Affinity Group at Indiana University.

Many thanks are also due my friends, including Margaret MacLaren, Julie Rolston, Belinda Edmondson, and Pablo DeGreiff, as well as Clare Rossini and Donna Chocol. Thanks to my mother, and to my father, for their support. And special thanks to Lisa Heldke for her encouragement and insight.

Prologue

Chaos theory is not as interesting as it sounds. How could it be? After all, the name "chaos theory" makes it seem as if science has discovered some new and definitive knowledge about utterly random and incomprehensible phenomena.

Actually, what seems to be going on is a kind of magic trick like the one Ludwig Wittgenstein described as putting something in a drawer and closing it, then turning around and opening the drawer, and removing the object with an expression of surprise. By calling certain physical systems "chaotic," scientists lead us to think that they are totally unintelligible—just a muddle of things happening with no connections or structures. So when they find interesting mathematical patterns in these unpredictable systems, they can exclaim that they have discovered the secrets of "order within chaos," even though only by christening these systems chaotic in the first place can they make such an impressive result possible.

Now there is nothing wrong with good marketing, especially when scientists need to present their work in attractive terms to publishers, journalists, funding agencies, and other scientists. The central insight of chaos theory—that systems governed by mathematically simple equations can exhibit elaborately complex, indeed unpredictable, behavior—is rightly seen as new and important. But sometimes there may be a temptation for researchers to hype their results, to make chaos theory sound too interesting, as if it will revolutionize our thinking not just about the physical world but about art and economics and religion as well. Considering that quantum mechanics was at least as revolutionary more than sixty years ago and that its cultural effects are still quite difficult to pin down, any grand claims for the implications of chaos theory may be setting us up for disappointment.

Nonetheless, the new conceptual approaches and experimental techniques used in chaos theory raise important philosophical questions about the meaning of the unpredictability manifested by the simple models studied and the nature of the scientific understanding they provide. These questions form the focus of this book, with each chapter devoted to one major issue. In more general terms, these separate arguments may be read as parts of a broader project, the goal of which is to demonstrate that the relatively new field of chaos theory is rich with philosophical interest.

I begin in chapter 1 with a brief discussion of just what chaos theory is, developing a provisional definition and presenting technical details and examples as required for the arguments that follow. Of course, any attempt to define chaos theory must confront the fact that most scientists and mathematicians rarely use the expression "chaos theory" at all, preferring to speak of "the study of chaotic phenomena" or "investigations of dynamical chaos." In this book I will use "chaos theory" to refer to these investigations, as have others who have sought to describe and understand this field of inquiry. But bear in mind that just as the "chaos" spoken of is not the chaos of incomprehensible tumult, the "theory" in chaos theory is not really a theory in the old-fashioned sense. There is no simple, powerful, and comprehensive theory of all chaotic phenomena, but rather a cluster of theoretical models, mathematical tools, and experimental techniques. While there is as yet no standard definition of chaos theory, I will concentrate on its crucial mathematical characteristic—its status as an application of dynamical systems theory—and its crucial scientific characteristic—its focus on unstable aperiodic behavior, which is intrinsically unpredictable.

This unpredictability results from the feature of all chaotic systems known as sensitive dependence on initial conditions.

This feature means that two systems that start very close together may eventually move very far apart, in much the same way that two paper cups floating next to each other at the top of a waterfall may well end up yards away from each other at the bottom. Even in a simple system, chaos means that if you are off by one part in a million, the error will become tremendously magnified in a short time. Sensitive dependence on initial conditions makes chaotic systems unpredictable because even the smallest degree of vagueness in specifying the initial state of the system will grow to confront the researcher with enormous errors in calculations of the system's future state.

This unpredictability provokes a chain of philosophical questions, the first of which is: Precisely what kind of limitation on scientific knowledge does sensitive dependence impose? In both special relativity and quantum mechanics, limitations on scientific knowledge are imposed because we have to revise basic physical concepts. Chaotic systems also present us with a limitation—namely, an intransigent unpredictability—yet they can appear in the context of exceedingly simple and entirely Newtonian equations of motion. If this limitation does not stem from some fundamental theoretical change, is it then merely a matter of practical difficulty?

After all, a chaotic system is predictable in principle: if we could exactly specify the initial situation, all future states would follow from straightforward calculation. The problem is that our initial specification must be impossibly accurate. For times far enough in the future, useful predictability would require an infinitely large device for storing and manipulating data. Is our inability to construct such a device really to be seen as a *practical* limitation? I contend in chapter 2 that here is a place where the line between "in theory" and "in practice" blurs. Sensitive dependence on initial conditions seems neither to challenge our fundamental theoretical beliefs nor to present merely an inconvenience for calculation; instead, chaos theory impels us to reassess the methodological assumption that small errors will stay small. Further, it invites us to reconsider our motivations for making that assumption.

The curious nature of chaotic unpredictability also raises questions about scientific determinism. These systems with sensitive dependence on initial conditions produce random or chancy behavior, the opposite of what we would expect from a picture of the universe as strictly determined clockwork. Yet chaos theorists consistently describe their models as "deterministic." In chapter 3 I elucidate the notions of determinism at work in chaos theory and explore the possible challenges posed by systems with sensitive dependence on initial conditions. Not only does chaos theory suggest that predictability should be considered separately from determinism, but in conjunction with quantum-mechanical considerations it raises serious doubts about the viability of the doctrine of determinism in the context of modern physics.

In chapter 4 I undertake to answer the broader question of what kind of understanding chaos theory provides. My conviction here is that recent developments in the sciences can help us refine or revise our philosophical conceptions of the character of scientific understanding. Specifically, chaos theory yields what I call "dynamic understanding"—a qualitative account of how order and unpredictability arise. The pursuit of dynamic understanding does not involve a search for exact quantitative predictions, and it eschews the use of microreductive explanations and rigorous deductive schemes. In spelling out the sense in which chaos theory may be said to be, for example, qualitative and holistic, I seek to avoid the twin perils of forcing chaos theory into preexisting theoretical models of science on the one hand or trumpeting it as a radically new and culturally superior form of science on the other.

The final chapter addresses some historical and sociological questions about the earlier history of nonlinear dynamics: why were chaotic systems not investigated until recently? I discuss some of the precursors of chaos, which lay unexplored for more than half a century, and consider some of the explanations that have been proposed to account for this situation. I conclude that a social and cultural interest in the exploitation of predictable natural processes must be part of an adequate explanation for the "nontreatment" of chaos. This final chapter represents both a sketch for a more detailed project in the history of science and an argument for the philosophical position that cultural factors can sometimes influence the choice of a science's mathematical tools.

In constructing my arguments, I will be following the common scientific practice of the persuasive use of citations to construct my case. While I am aware of controversy surrounding many of the points I raise, I am not presenting this work as an exercise in systematic philosophy that will use chaos theory as an example of how to unify the views of all the figures I cite. Instead, my intent is to use the work of a variety of philosophers in order to explore the issues arising from the scientific study of chaos.

It would be inappropriate to fail to mention some of the philosophical aspects of chaos theory that will not be dealt with in this book. First among these is the issue of chaos theory and randomness, first broached in chapter 1. I have not attempted to give a rigorous account of how deterministic dynamical systems can be said to be random or to assess the implications that chaos theory has for our very definition of randomness. As I note in chapter 3, I have also passed over the issue of chaos theory and the determinism–free will controversy, although some researchers have made strong claims that chaos provides a way to resolve this problem. Finally, while the bizarrely beautiful figures known as fractals do play a role in the study of chaotic dynamical systems, I have not undertaken an examination of the implications fractal geometry may have for the philosophy of mathematics.

Despite these limitations, the aspects of chaos theory presented here suggest that this field of scientific research promises much fruitful philosophical investigation. As a new area of inquiry, the scientific study of chaos provides an occasion for investigating the interaction between our methods for gaining knowledge about the world, our notions of what that knowledge should look like, and our conceptions of what kind of world we inhabit. My goal in this book is to begin to explore this interaction between methodology, epistemology, and metaphysics in the context of nonlinear dynamics and chaos.

1

What Is Chaos Theory?

> The world must actually be such as to generate ignorance and inquiry; doubt and hypothesis, trial and temporal conclusions. . . . The ultimate evidence of genuine hazard, contingency, irregularity and indeterminateness in nature is thus found in the occurrence of thinking.
>
> —John Dewey (1958)

In the past 25 years, scientists working in such fields as fluid mechanics, chemistry, and population biology have developed many successful mathematical models for natural phenomena. Some of these models have two features that seem highly incongruous: they consist of only a few simple equations, yet the solutions to these equations display extremely complicated and even unpredictable behavior. The same intricate patterns are found in vastly different realms of investigation, and apparently trivial computer programs can accurately simulate these patterns amidst otherwise random sequences of digits. The analysis of these models and the investigation of similar behavior in actual experiments have been termed "chaos theory."

A Provisional Definition

Chaos theory is a young field of scientific inquiry that stretches across many established disciplines, blurring old distinctions and creating new ones. Constructing a definition can aid in sketching the domain of chaos theory, by suggesting the kinds of questions it asks and the kinds of physical systems about which it asks them. I suggest the following definition: chaos theory is *the qualitative study of unstable aperiodic behavior in deterministic nonlinear dynamical systems.*

Let us begin at the end of this definition and consider the respects in which chaos theory is an investigation of dynamical systems. When scientists confine attention to a particular collection of objects or processes, they draw a figurative frame around the subject matter of their inquiry and label the contents of that frame a "system." A system may be the sun and its satellites or the population of one species of turtle on an island, a beating heart or one muscle cell in a petri dish.

Some aspects of the system are deemed scientifically relevant, and some of these aspects admit of mathematical description. By specifying the numerical value of all quantitative features of a system, one obtains a compact description of the way the system is at a certain time. Such a description, given in terms of the values of one or more variables, is termed the "state" of the system at that time. A *dynamical system* includes both a recipe for producing such a mathematical description of the instantaneous state of a physical system and a rule for transforming the current state description into a description for some future, or perhaps past, time. A dynamical system is thus a simplified model for the time-varying behavior of an actual system.

The most common types of dynamical systems are differentiable ones, where the relevant variables change in a smooth or continuous way. In this case, the rules governing the changing state of the system (the "evolution equations") can usually be written in the form of differential equations that specify the rates of change of the variables. Sometimes the system is best described in terms of discrete time units, as is the case when we are interested in the total number of cases of a particular disease each year. In such a circumstance, the evolution equations take the form of a difference equation or mapping.

Given the state of a dynamical system at one particular time and the evolution equations, it is possible to calculate the state at other times by using these differential equations. By changing the system's variables in small increments, one can discover how the state of the system changes from the initial time to the final time. This procedure has the disadvantage of often becoming computationally unwieldy—such an "open-form" solution to the evolution equations can require burdensome calculations. Some differentiable dynamical systems can be manipulated by straightforward mathematical techniques to yield a closed-form solution. Such a solution typically takes the shape of a simple formula that allows one to bypass the evolution equations. Instead of requiring one to recalculate the state of the system for each increment between the initial time and the final time, we can simply plug the final time into the formula and find the final state of the system.[1]

One distinguishing mark of the dynamical systems of interest for chaos theory is the presence of *nonlinear* terms in the equations. Nonlinear terms involve algebraic or other more complicated functions of the system variables. For example, in a system with two variables x and y, expressions such as x^2 or $\sin(x)$ or $5xy$ would be nonlinear terms. These terms may stem from the inclusion of such factors as frictional forces or limits to biological populations. The nonlinearity of the equations usually renders a closed-form solution impossible. So researchers into chaotic phenomena seek a *qualitative* account of the behavior of nonlinear differentiable dynamical systems. After modeling a physical system with a set of equations, they do not concentrate on finding a formula that will make possible the exact prediction of a future state from a present one. Instead, they use mathematical techniques to "provide some idea about the long-term behavior of the solutions" (Devaney 1986, 4).

As a qualitative study, chaos theory investigates a system by asking about the general character of its long-term behavior, rather than seeking to arrive at numerical predictions

1. Although very few problems allow a closed-form solution, this approach is presumed to be the norm for most physical sciences.

about its exact future state. A closed-form solution may allow one to predict, for example, when three planets traveling in elliptical orbits will line up, whereas a qualitative study would be more interested in discovering what circumstances will lead to elliptical orbits as opposed to, say, circular or parabolic ones.[2] The investigation of the qualitative aspects of a system's behavior began with the work of Henri Poincaré. Current research in this field goes by the name "dynamical systems theory," and it typically asks such questions as, what characteristics will *all* solutions of this system ultimately exhibit? And how does this system change from exhibiting one kind of behavior to another kind? Chaos theory is a specialized application of dynamical systems theory.

While qualitative questions can be asked about almost any dynamical system, chaos theory focuses on certain forms of behavior—behavior which is *unstable* and *aperiodic*. Instability will be discussed in more detail below, but for now it is sufficient to mention that unstable behavior means that the system never settles into a form of behavior that resists small disturbances. A system marked by stability, on the other hand, will shrug off a small jostle and continue about its business like a marble which, when jarred, will come again to rest at the bottom of a bowl, or a watch, which will continue ticking reliably after receiving a slight jolt.

Aperiodic behavior occurs when no variable describing the state of the system undergoes a regular repetition of values. Unstable aperiodic behavior is highly complex: it never repeats and it continues to manifest the effects of any small perturbation. Such behavior makes exact predictions impossible and produces a series of measurements that appear random.

Behavior that is both unstable and aperiodic may be initially hard to picture, but the difficulty disappears if we consider, for example, human history. Although broad patterns

2. Describing these investigations as qualitative may obscure the fact that the research is still technical and mathematical in nature. As described below and again later in chap. 4, chaos theory often yields precise numerical results.

in the rise and fall of civilizations may be sketched, events never repeat exactly—history is aperiodic. And history books teem with examples of small events that led to momentous and long-lasting changes in the course of human affairs. The standard examples of unstable aperiodic behavior have always involved huge conglomerations of interacting units. The systems may be composed of competing human agents or colliding gas molecules, but until recently our primary image of behavior so complex as to be unstable and aperiodic was the image of a crowd.

One last determinative must be added to our definition of chaos theory to mark the fact that it represents a break with traditional accounts of complex behavior in physical systems. A distinguishing feature of the systems studied by chaos theory, and a large part of what makes the field so exciting to researchers, is that unstable aperiodic behavior can be found in mathematically simple systems.[3] These systems bear the label *deterministic* because they are composed of only a few (typically less than five) differential or difference equations, and because the equations make no explicit reference to chance mechanisms. We are accustomed to thinking that behavior that shows no repeating patterns and responds abruptly to even small disturbances must result from the competing and reinforcing influences of countless subsystems. But chaos theory explores very simple, rigorously defined mathematical models that nonetheless display behavior so complex as to merit description as random.

Defining chaos theory as the qualitative study of unstable aperiodic behavior in deterministic nonlinear dynamical systems presents the apparent paradox of finding complex and even random behavior in very simple and ordered systems. To explore the reasons why chaos theory seems to trade in such contradictions, it will help to examine a concrete example of

3. Furthermore, chaos is an appropriate label only when such behavior occurs in a bounded system. An explosion does not qualify as chaotic behavior in this sense.

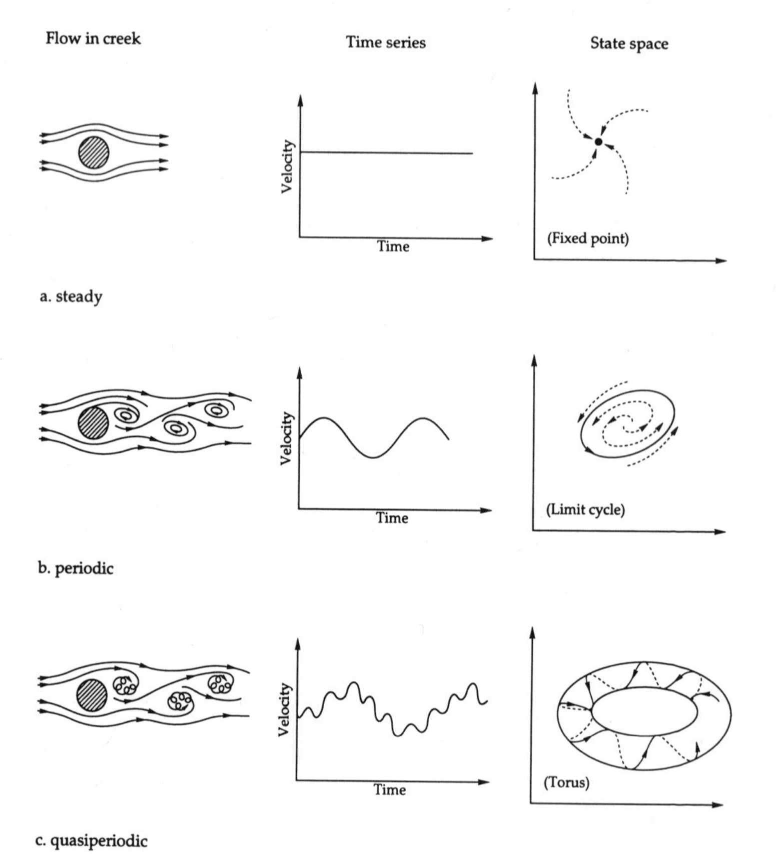

Fig. 1. An early transition to turbulence. After Kadanoff 1983

how "orderly disorder" provided a crucial clue to a long-standing problem in classical physics: the problem of the onset of turbulence.

Two Pictures of Complexity

Turbulence remains an unsolved problem for classical physics; there is still no adequate theoretical account of the whorls and eddies that appear in waterfalls, whirlpools, and wakes. One of the few theoretical approaches to this problem before chaos theory was the account suggested by Lev Landau (1944). The Landau model seeks to understand turbulence by describing how smooth fluid flow becomes disrupted as the speed (for example) of the fluid past an object is increased. By understanding how turbulence begins, it is hoped that some clues can be found to the nature of full-blown turbulent behavior.[4]

Imagine a slowly moving creek in which water flows past a large rock, and a very accurate device downstream from the rock that measures the velocity of the water at one point. For steady flow, the device will register a constant value, but as the water increases in speed, the smooth flow lines around the rock begin to bend, causing undulations that detach into small eddies that move downstream in time. As one of these eddies passes our measuring device, the velocity will register an increase, then a decrease, and then return to the undisturbed value until another eddy passes by. The sequence of velocity measurements—the time series—changes from constant to periodic behavior (see fig. 1a–b).

Perhaps the most important arena for understanding dynamical systems is state space (sometimes called "phase space"), a mathematically constructed conceptual space where each dimension corresponds to one variable of the system. Thus, every point in state space represents a full description of the system in one of its possible states, and the evolution of

4. See the discussion in Kellert, Stone, and Fine 1990.

the system manifests itself as the tracing out of a path, or trajectory, in state space. This method is extremely useful, because it is often possible to study the geometric features of these trajectories without explicit knowledge of the solutions they represent. The characterization of possible trajectories in state space according to their "shape"—a kind of topological taxonomy—constitutes a major element of the mathematical apparatus of chaos theory.

Consider a mathematical representation of all possible states of the creek—every point in this state space corresponds to a different configuration of the fluid flow. In the case of steady flow, the behavior of the system is characterized by a single attracting point, which remains motionless. No matter where you begin in the state space, no matter how you stir up the creek as you begin its flow at this speed, the system will eventually wind up moving along smoothly with a constant velocity everywhere in the creek bed. The transition to the "small eddies" behavior has a mathematical counterpart in the change from an attractive point to an attractive periodic cycle in state space (see fig. 1b). Such a change in the nature of a system's behavior as a parameter is varied is termed a bifurcation.

As the speed of the creek increases, the behavior of the flow becomes more complicated and may depend on the configuration of the creek bed and the obstacle. But we can follow one scenario for the transition to turbulence, where the faster flow makes smaller eddies appear within the original ones. The time series of velocity measurements will now vary with two frequencies (see fig. 1c). The system has undergone another bifurcation and the point representing the state of the system in abstract state space now spirals around a two-dimensional torus, or doughnut shape. If the two frequencies are not rationally commensurate, the system will never return to exactly the same state—it will wind around and around the torus forever and never return to its starting point. This situation is labeled quasiperiodic behavior and can best be appreciated by imagining a clock with two hands, one of which circles the clock face every hour and one of which takes π hours to describe a circle. If we start the clock with both hands at twelve, they will never again meet at twelve; such a system is deterministic and easily predictable, but not periodic.

The heart of the Landau model's explanation of the onset of turbulence is this: as the flow rate is increased, the quasiperiodic motion on the two-dimensional torus becomes unstable; some small disturbance will lead to three-dimensional quasiperiodic motion, then four-dimensional, and so on to infinity. The onset of turbulence represents the piling up of huge numbers of incommensurable frequencies, representing the excitation of more and more degrees of freedom—more and more eddies within eddies. Quasiperiodic motion on a very high dimensional torus (and, in the limit, a torus of infinite dimensions) will never repeat itself and will be utterly unpredictable because of a huge number of competing influences. So the Landau model accounts for the characteristic aperiodicity and unpredictability of turbulence. It thus suggests that complex, apparently random turbulent behavior is best understood as akin to a clockmaker's shop with a huge number of clocks each ticking at a different, irrational rhythm.

One of the birthplaces of chaos theory was in an alternative account for the onset of turbulence, an account that challenged this picture of complexity. Known as the Ruelle-Takens-Newhouse (RTN) model, this account rejects the idea that the complex behavior in turbulent flow must be modeled by the unwieldy agglomeration of incommensurable frequencies. The transition to turbulence is explained instead by the appearance in state space of an attractor that represents very complicated dynamical behavior, yet is described by a very simple set of mathematical equations. Such a novel mathematical object is called a "strange attractor."

In the RTN model, the behavior of fluid flow past an obstacle follows the path laid down by Landau only up to the appearance of a two-dimensional torus. After that point, a further increase in the flow rate can render this attractor unstable, and for a wide variety of cases the behavior will change to weak turbulence characterized by motion on a strange attractor (Ruelle and Takens 1971; Newhouse, Ruelle, and Takens 1978). The strange attractor has several important characteristics that will be discussed more fully in the next section: (1) it is an attractor, that is, an object with no volume in state space toward which all nearby trajectories will converge; (2) it typically has the appearance of a fractal, that is, a stack of two-dimensional sheets displaying a self-similar packing structure; (3) motion on it exhibits a form of instability known as sensitive dependence on initial conditions, that is, for any point on the attractor there is another point nearby that will follow a path diverging exponentially from the path of the first; (4) it can be generated by the numerical integration of a very simple set of dynamical equations.

The idea that complex and unpredictable behavior such as turbulence can be understood by investigating simple dynamical systems is a bold hypothesis at the heart of chaos theory. The RTN model made this hypothesis mathematically plausible, and the work of other mathematicians, theoreticians, and experimenters in the seventies added evidence of the fruitfulness of chaos theory. Edward Lorenz laid the groundwork for this approach with his discovery of a strange attractor in a highly simplified set of equations derived from a model for fluid convection (Lorenz 1963). The strange attractor that bears his name provides an appropriate place to begin a fuller discussion of the mathematical theory of chaos.

Strange Attractors

The first picture of a strange attractor appears in Edward Lorenz's paper "Deterministic Aperiodic Flow" (1963). As a meteorologist, Lorenz was interested in mathematical models for the behavior of the earth's atmosphere. As a student of the mathematical physicist David Birkhoff, he followed the dynamical systems approach favored by his teacher. Lorenz started with the Navier-Stokes equations, a powerful but difficult to solve mathematical description for the behavior of incompressible fluids. Building on the work of Saltzman (1962), he simplified the equations by truncating them—paring them down to a dynamical system consisting of these three ordinary differential equations:

$$\dot{x} = -\sigma x + \sigma y$$
$$\dot{y} = -xz + rx - y$$
$$\dot{z} = xy - bz$$

In this system, the Lorenz system, the state of the model's "atmosphere" is entirely specified by the variables x, y, and z. The characteristics of the system can be varied by altering the parameters σ, r and b, which would correspond to changing certain physical properties of the model's "air."[5] The system is deterministic and mathematically straightforward, although it does not admit of a closed-form solution. To understand the behavior of this system, Lorenz made use of a relatively new device, the digital computer, which could undertake the laborious task of integrating the equations step by step in order to construct a typical solution to the system.[6]

For certain values of the parameters, the solutions of the Lorenz system display a peculiar type of instability. If you specify the initial state of the system by stipulating initial values x_0, y_0, and z_0, you will obtain a solution corresponding to fluid convection that rotates first clockwise for a while, then counterclockwise for a different amount of time, then back to clockwise, and so on. The behavior is quite complicated but is entirely governed by the differential equations. However, if you were to start with a slightly different trio of initial condi-

5. For example, the parameter σ represents the Prandtl number: kinetic viscosity divided by thermal diffusivity.

6. Starting with some trio of values for x, y, and z, the computer advances "time" in the model in small increments, slowly building up a series of later states of the system. This series describes a path in the state space and constitutes one solution of the dynamical system. Note that this solution is not a *general* solution in the sense of a closed-form solution.

tions, say $x_0 + 0.001$, y_0, and z_0, the solution would become substantially different in only a short time. The behavior would still be rotational convection first in one direction then another, but the states of the two systems, which initially were extremely similar, will rapidly diverge.

This form of instability bears the name *sensitive dependence on initial conditions*, and is a distinguishing characteristic of chaotic behavior. A dynamical system that exhibits sensitive dependence on initial conditions will produce markedly different solutions for two specifications of initial states that are initially very close together. In fact, given any specification of initial conditions, there is another set of initial conditions close to it that will diverge from it by some required distance, given enough time.[7]

Lorenz spelled out the consequences of his discovery as follows: "It implies that two states differing by imperceptible amounts may eventually evolve into two considerably different states. If, then, there is any error whatever in observing the present state—and in any real system such errors seem inevitable—an acceptable prediction of an instantaneous state in the distant future may well be impossible" (1963, 133). Systems that display sensitive dependence on initial conditions exhibit what Lorenz labeled the "butterfly effect."[8] If we stipulate that the earth's weather is a chaotic system, then the flap of a butterfly's wings in Brazil today may make the difference between calm weather and a tornado in Texas next month. Any attempt to predict the weather with long-term precision would fail utterly unless it took into account all

7. A formal definition for sensitive dependence on initial conditions in the one-dimensional case is given by Devaney (1986, 49): a function f on an interval J has sensitive dependence on initial conditions "if there exists $\delta > 0$ such that, for any $x \in J$ and any neighborhood N of x, there exists $y \in N$ and $n \geq 0$ such that $|f^n(x) - f^n(y)| > \delta$."

8. James Gleick traces the origin of this phrase to a paper Lorenz delivered at the 1979 annual meeting of the American Association for the Advancement of Science. The paper was entitled, "Predictability: Does the Flap of a Butterfly's Wings in Brazil Set Off a Tornado in Texas?"

data, including all butterflies, with complete accuracy.[9] Chaotic dynamical systems modeled on computers, for instance, may transform a round-off error at the limit of the machine's precision (the twentieth decimal place, for example, in a typical double-precision calculation) into a drastic divergence in the state of the system. Such extreme sensitivity makes computation of exact trajectories impossible.

In the face of such impossibility, Lorenz used a qualitative approach to study his system. The first step in such an approach is to note that the system is *dissipative,* meaning that friction causes a loss of available energy: the system would wind down to motionlessness if the external source of energy—in this case, the heat that causes convection—were removed. The state space representation of the behavior of a dissipative system displays a contraction of areas. The end result is an attractor—a set of points such that all trajectories nearby converge to it.[10] Until the advent of chaos theory, only three types of attractors were generally recognized: the fixed point, the limit cycle, and the torus. But none of these attractors can describe the unstable aperiodic motion Lorenz found. By using a computer to plot the trajectory of his system, Lorenz created the first picture of a surprising new geometrical object: a strange attractor (see fig. 2).

Part of the reason these objects are called strange is that they reconcile two seemingly contradictory effects: they are attractors, which means that nearby trajectories converge onto them, and they exhibit sensitive dependence on initial conditions, which means that trajectories initially close to-

9. The "butterfly effect" is sometimes interpreted in a way that suggests that the careless insect actually causes the tornado. As I discuss in chap. 4, this causal language is inappropriate.

10. For a dissipative dynamical system an attractor may be defined as: a set of zero volume in state space which is invariant under the action of the evolution equations and which is surrounded by a domain of nonzero volume such that any trajectory which originates within that domain converges asymptotically onto the attractor (Bergé, Pomeau, and Vidal 1984, 111). The geometrical properties of the attractor thus characterize the qualitative properties of the dynamical system once any transient effects have died away.

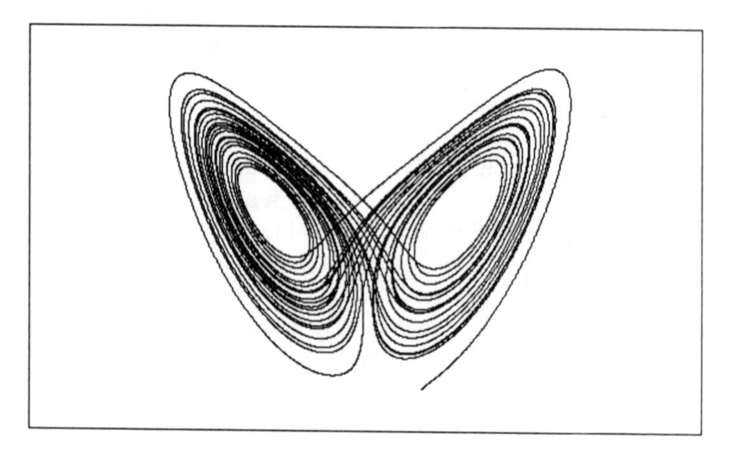

Fig. 2. The Lorenz attractor

gether on the attractor diverge rapidly. This apparent contradiction is reconciled by one of the main geometric features of strange attractors: a combination of stretching and folding. The action of a chaotic system will take nearby points and stretch them apart in a certain direction, thus creating the local divergence responsible for unpredictability. But the system also acts to "fold" together points that are at some distance, causing a convergence of trajectories in a different direction.

In the Lorenz attractor, we see two disklike structures, one corresponding roughly to a clockwise sense of rotational convection and the other to a counterclockwise sense. As two nearby points on the left "disk" evolve in time, their trajectories move toward the center of the figure and are stretched apart. After a few trips around the left disk, one of the trajectories may have wandered far enough so that on its next pass through the center it will split off from its former neighbor and begin spinning around the right disk. And all the time, trajectories from the right disk are being "folded" in among those on the left disk as well. Thus we see the two attributes of a strange attractor: nearby points can quickly evolve to opposite sides of the attractor, yet the trajectories are confined to a region of phase space with a particular shape.

That shape is a fractal. The stretching and folding of chaotic systems gives strange attractors the distinguishing characteristic of a nonintegral, or fractal, dimension. In *Order within Chaos* we find the following argument for why chaos theory finds itself linked with the study of these now-notorious shapes.[11] As Lorenz showed, a dissipative system in three variables can behave chaotically. As a dissipative system, any specified area of initial conditions will contract with time. The attractor represents the shape that any initial set of points will approach asymptotically, so it must have no volume in the three-dimensional state space. Therefore, the dimension of the attractor must be less than three. But a two-dimensional surface such as a sphere or a torus cannot be the attractor, because nearby trajectories cannot diverge on such a surface without intersecting.

Sensitive dependence on initial conditions requires that all points on the surface have nearby points that diverge exponentially, but the topological facts of a two-dimensional surface dictate that this would lead to trajectories that cross. Trajectories must not cross because that would contradict the deterministic nature of the system of differential equations: intersecting trajectories would mean that the system would have a "choice" of paths when it reached the intersection. Thus, the attractor must have a dimension greater than two (Bergé, Pomeau, and Vidal 1984, 122). The Lorenz attractor would therefore have to have a dimension between 2 and 3. A geometric object with nonintegral dimension is called a fractal.

A jagged coastline provides a useful example of a fractal-like object. Observed from afar, the coastline reveals some peninsulas and bays; on closer examination, smaller juts and coves are seen, and these again reveal jagged borders when

11. *Order within Chaos: Towards a Deterministic Approach to Turbulence* (Bergé, Pomeau, and Vidal 1984) provides an excellent introduction to the main mathematical, physical, and experimental features of research into dynamical chaos.

surveyed more closely. If we imagine the coastline so jagged that with each new level of magnification new details of the terrain appear, so that the line describing the coast began to "take up space," this is a fractal. In the case of a strange attractor, the stretching and refolding action of chaotic dynamical systems often produces objects with the appearance of infinite puff pastry—stacks of sheets that are themselves two-dimensional, but stacked in a never-ending self-similar structure that seems to intrude into the three-dimensional space. The dimension of such an object, conceived as a measure of its "intrusiveness," is more than 2 but less than 3.

Another exemplary strange attractor is the Hénon attractor, which appears to be a line stretched and folded so that it begins to have nonzero area. This attractor is produced by the Hénon mapping, a discrete dynamical system that represents the simplest nonlinear two-dimensional system with a constant rate of attraction (Hénon 1976, 71–72):

$$x_{n+1} = y_n + 1 - ax_n^2$$
$$y_{n+1} = bx_n$$

When $a = 1.4$ and $b = 0.3$, the attractor for this system looks like figure 3.

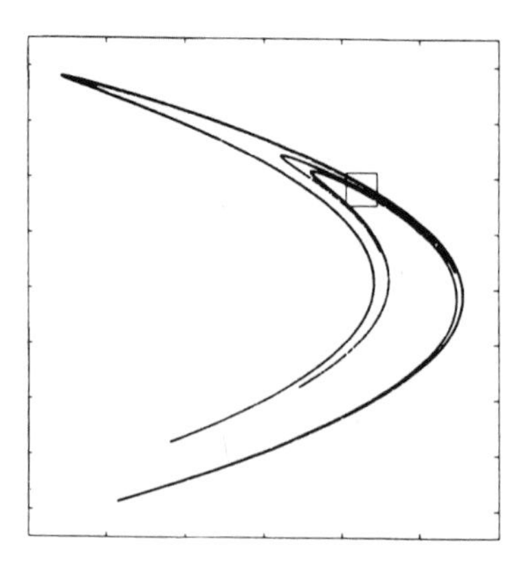

Fig. 3. The Hénon attractor. From Hénon 1976

This mapping has a smaller contraction rate than the Lorenz system, meaning its layers are not folded together too closely to be discerned (Hénon 1976, 281). As figure 4 illustrates, successive enlargements of the Hénon attractor reveal its self-similar fractal structure. The Hénon attractor has a dimension between 1 and 2.

The nonintegral dimension of strange attractors is more than a geometric curiosity, however, for this dimension can be measured and thus provides a useful topological characterization of the behavior of the system. The possibility of such measurement rests on a definition of dimension different from the usual Euclidean definition. If we seek to measure the fractal dimension of an attractor in *n*-dimensional space, we can start by covering it with *n*-dimensional cubes (or hypercubes) of a relatively large size. After counting how many cubes it takes to cover the object, we reduce the cubes' edge-length, *e*, and count how many of these smaller cubes are needed. This measure of fractal dimension, known as the Hausdorff-Besicovitch dimension *D*, is defined by $D = \lim_{e \to 0} (\log N(e))/\log(1/e)$, where $N(e)$ is the number of cubes of edge-length *e* needed to cover the attractor (Froehling et al. 1981, 607).

Roughly speaking, the fractal dimension describes the scaling properties of a geometrical object, characterizing the way the structure of the object reappears at different degrees of "magnification." By characterizing how tightly the layers within the attractor are packed, a purely topological feature of the attractor gives a quantitative indication of the stretching and folding at work in the dynamical system. There are several definitions of fractal dimension that researchers find useful for different purposes, but all these quantitative measures allow researchers to investigate the way a system changes its behavior in response to a change in the parameters describing the system and its environment.[12]

Another quantitative characterization of chaotic systems is

12. Some alternative formulations of fractal dimension are given in Grassberger and Procaccia 1983 and Mayer-Kress 1986.

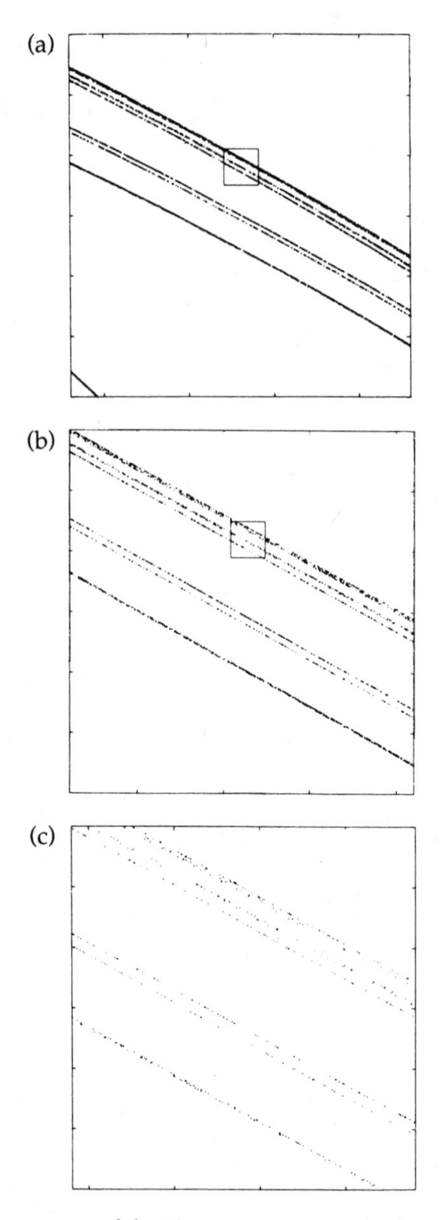

Fig. 4. The structure of the Hénon attractor. (a) Enlargement of the boxed region in fig. 3. (b) Enlargement of the boxed region in (a). (c). Enlargement of the boxed region in (b). From Hénon 1976

given by the Lyapunov exponents, which measure the degree of sensitivity to initial conditions and thus the degree of unpredictability. Determination of the Lyapunov exponents for a system proceeds by constructing a small "ball" of initial conditions around a certain point in state space, where each point on the ball represents a small displacement from the central point. The evolution equations will act on this ball to change its shape: the ball will be stretched out or compressed in each direction depending on whether the dynamical system acts to magnify or reduce small displacements. For a three-dimensional system, the deformation of the ball can be approximately described by three numbers, corresponding to the deformation along the trajectory of the central point and along two directions perpendicular to it. These three numbers are the Lyapunov exponents of the system (see Shimada and Nagashima 1979, 1613; Bergé, Pomeau, and Vidal 1984, 280).

If the Lyapunov exponent in a certain direction is negative, any small displacement in that direction will shrink exponentially as the system evolves. If the exponent is zero, the displacement will remain roughly the same in magnitude. But if the Lyapunov exponent has a positive value, any small displacement will grow exponentially in time (at least until the large-scale folding action of the attractor takes over). The presence of a positive Lyapunov exponent signals sensitive dependence on initial conditions, and its size provides a measure of how quickly an uncertainty about initial conditions will grow to make predictions impossible. Measurement of the largest Lyapunov exponent of a system supplies a sign of the presence of chaos and a way to quantify its strength.

The measurement of Lyapunov exponents can be used even when the trajectories of a dynamical system do not lie on an attractor: one chooses a trajectory as a standard and measures the growth or shrinking of small displacements from it. It should be noted that strange attractors appear only in chaotic dissipative systems. For Hamiltonian systems, where energy is conserved, there is no convergence onto an attractor. Instead, trajectories are confined to a surface of constant energy. Chaotic behavior can occur in such systems, but instead of strange attractors with fractal dimension, the trajectories will fill the allowed energy surface (which may itself display an extremely complicated structure).[13]

The Transition to Chaos

In conservative systems, as in dissipative ones, much attention is focused on the way a system changes from ordinary behavior to chaos. The strange attractors just discussed promise a new approach to the study of the onset of turbulence by supplying an alternative model for complex and unpredictable behavior. Another simple dynamical system, the logistic map, has been used to model the transition to chaos.

The logistic map arose from research into the annual fluctuations of insects like the gypsy moth, which breed and then die off en masse, producing a new nonoverlapping generation each year. In modeling the population of these insects, a number between 0 and 1 indicates the state of the system—the insect population for each year: zero represents total extinction and 1 represents the largest possible number of gypsy moths. If we know X_{1993}, the number corresponding to the insect abundance this year, then our simple model will tell us what X_{1994} will be. This model, the logistic map, takes the form of the evolution equation $X_{n+1} = \alpha x_n(1-x_n)$, where α is a parameter. This parameter dictates how drastically the insect population responds to favorable or unfavorable conditions—the higher α is, the more rapidly will a small population grow or a large population diminish. By repeatedly applying the evolution equation, we arrive at a series of yearly values for X that may behave quite differently depending on the value chosen for α.

For some values of α, the dynamical system will reach equilibrium: whatever the initial population, succeeding years will find the insects settling down to a stable number that precisely

13. See Walker and Ford 1969, Chirikov 1979, and Moon 1987, 63.

reproduces itself year after year. As α increases, we find that the successive values of X will converge to a two-year cycle: a large number of insects produces a smaller population the next year, which in turn reproduces the original large number of insects the following year and so on. Further tinkering reveals that for a higher value of α we can find a cycle that repeats every four years. Even higher values display a cycle with twice that period, repeating every 8 years, then every 16, 32, and so on, in what is called a "period-doubling cascade." Finally, for a high enough value of α the behavior becomes aperiodic and indeed chaotic: no matter what the initial number of insects is, the sequence of numbers that follows from the formula will never repeat.

The examination of this very simple nonlinear dynamical system has yielded a tremendously rich variety of behaviors, most notable the period-doubling cascade. Indeed this system, which in some ways is the simplest possible nonlinear difference equation in one variable, has become one paradigm for the whole of chaos theory (Moon 1987, 67). One reason for this is that the period-doubling bifurcations obey some surprising scaling relations. The values of the parameter α at which each successive cycle first appears grow closer and closer together in a manner converging on a constant ratio. If α_k is the value at which the 2^k-cycle first appears, then the limiting value as k goes to infinity of the quantity $(\alpha_k - \alpha_{k-1})/$ $(\alpha_{k+1} - \alpha_k$ is $\delta = 4.669201\ldots$, a value found by repeated computer experimentation. What is especially surprising is that period-doubling behavior is found in a tremendous variety of nonlinear systems. For any continuous differentiable function having an extremum (a "bump") between 0 and 1 such that the function near the extremum is roughly quadratic in form, period-doubling behavior will be observed and it will display the same scaling features with the same constants as the logistic map (see Feigenbaum 1978; see also Lanford 1982).

This result allows us to make quantitative predictions about a system's period-doubling behavior as soon as we have verified that certain qualitative conditions hold (Bergé, Pomeau, and Vidal 1984, 201). For instance, even without knowledge of the exact equation governing a system, we can tell at what value of the adjustable parameter the next bifurcation will occur, or when the threshold to chaos will be crossed. In this way, the study of one-dimensional mappings can produce results applicable to systems in several dimensions; once period-doubling is observed in a system, powerful mathematical tools can be brought to bear in order to understand how the system makes the transition to chaos.

In the logistic map, full-blown chaos occurs at roughly a value of $\alpha = 3.57\ldots$, and the chaotic behavior manifests the greatest sensitive dependence on initial conditions at the maximum value of $\alpha = 4$. Yet even here, every number in the series generated by the dynamical system follows from its predecessor according to a strict rule. How then can such a sequence be called random? Researchers in chaos theory usually answer this question by appealing to an operational sense of randomness as utter unpredictability.

The sensitive dependence on initial conditions in the logistic map is so great when $\alpha = 4$ that each discrete unit of time "uses up" one binary digit of information about the initial state of the system. Since any specification of the initial state will be only finitely accurate, there will come a time when the residual inaccuracy will be so magnified as to drown out all the information contained in the initial specification. Even such minute details as the round-off algorithm of the calculation device used will come to dictate the trajectory the system follows. Once such a situation occurs, the series of numbers generated by the evolution equations no longer bears any relationship to the initial state specified: all predictability is lost and the dynamical system can be treated as if it were random.

Besides the period-doubling route to chaos, there are two other transitional scenarios that have received some theoretical attention: the transitions via quasiperiodicity and intermittency. Quasiperiodicity is the route described in the RTN model, where a torus in state space changes into a strange attractor. Intermittency occurs when a periodic signal is interrupted by random bursts that arrive unpredictably but increasingly often as a parameter is increased. It should be noted that while chaos theory permits qualitative understanding and even some quantitative prediction with regard to these three routes, no one has yet established necessary or sufficient conditions for determining which type of transition will occur in a given system. But when a system manifests aspects of a certain type of transition, the mathematical theory pertaining to that generic type can be applied.

Finding Chaos in Experimental Systems

The qualitative study of chaotic dynamical systems is mathematically interesting, but how can it be of use in experimental situations in which we do not know the equations governing the physical system's behavior? Several experimental techniques have been devised to allow researchers to discern strange attractors in just these situations, and also to measure fractal dimensions and Lyapunov exponents as well.

In general, the first step in an experimental study of a possibly chaotic system is to obtain a detailed record of the value of one variable of the system. Simple inspection of this time-series can reveal behavior that is obviously periodic, intermittent, or irregular. But an important initial analysis comes from the technique of transforming the time-series data into what is known as a Fourier power spectrum (see, for example, Bergé, Pomeau, and Vidal 1984, chap. 3). This spectrum reveals a sharp peak at the frequency of each periodic component in the data stream. The spectrum for a system following a stable limit cycle will thus have one strong peak, while quasiperiodic behavior will display several peaks and their harmonics, and chaotic regimes produce a broad band. A quasiperiodic regime with a great number of frequencies present will also seem to have a broad band power spectrum, however, so this method is not a fail-safe way to detect deterministic chaos as opposed to Landau-style noise. But it can serve to display the successive appearance of halved frequencies during period-doubling, for instance.

One of the most important methods for discovering and analyzing chaos in dissipative systems is the reconstruction of attractors, a procedure that extracts the geometric features of a system's behavior from the time-series record. This method, developed by the physicists N. Packard, J. Crutchfield, J. Farmer, and R. Shaw together with the mathematical work of Floris Takens (1981), allows researchers to study qualitative features without solving (or even knowing) the equations governing a system. The basic idea is to reconstruct a multidimensional attractor from the time series by plotting, say, $x(t)$ versus $x(t+\tau)$ and $x(t + 2\tau)$, where τ is a suitable time-lag (Packard et al. 1980, 713). Thus, for the three-dimensional case, three measurements of the same variable serve as three independent variables in order to specify the state of the system. As they write, "the evolution of any single component of a system is determined by the other components with which it interacts. Information about the relevant components is thus implicitly contained in the history of any single component" (Crutchfield et al. 1986, 54).

The reconstruction of attractors creates a simulated state space out of the one-dimensional time-series record. Because the most important properties of strange attractors are topological, almost any set of coordinates can be used to discern these properties, so this method is usually a reliable guide to the dynamical behavior of the system under study (Shaw 1981b, 222). Numerical simulations have confirmed that reconstructing a three-dimensional attractor from $x(t)$, $x(t + \tau)$, and $x(t + 2\tau)$ yields a representation in the constructed state space that is "faithful" to "the dynamics in the original x, y, z space" (Packard et al. 1980, 714).[14]

14. This method should not be considered universally applicable, however. For example, one would not expect to get a low-dimensional attractor from data collected at one point of a very large, spatially extended system with many degrees of freedom (like the earth's atmosphere). In questionable situations, one may need to collect additional data, or to achieve greater ac-

Once a reconstructed picture of the dynamics is available, researchers may wish to determine the dimension of the attractor. The fractal dimension may be computed by various techniques that take off from the Hausdorff-Besicovitch definition of topological dimension, including an analysis of the density of points on the attractor within spheres of increasing size (Moon 1987, 220). The development of more efficient ways to calculate the dimension of attractors, and the invention of newer, even more informative quantitative measures for their topological features, attract a tremendous amount of interest among those currently working in chaos theory.

Another useful analytical tool for studying reconstructed attractors with few degrees of freedom is the Poincaré surface-of-section. This method involves examining the reconstructed trajectories of a system as they pass through a plane in state space. Imagine a very thin phosphorescent screen that slices the attractor. Instead of trying to visualize the attractor itself, the surface-of-section allows us to pay attention only to the pattern of glowing spots where the trajectories intersect the screen. Since a two-dimensional display is easier to examine, these surfaces-of-section are often used to look for the characteristic doubling of paths on the period-doubling route to chaos or for the intricate folded structure that often signals a chaotic system (Moon 1987, 53).

If the system is highly dissipative, the surface-of-section will appear to be very thin—practically a line segment. In this case, another analytic technique is used to discover whether the system is chaotic or "merely stochastic." If such a system were stochastic—meaning, not governed by a few deterministic equations—we would expect the trajectory to be randomized each time it passed through the thin segment on the plane. So we plot the position along the segment versus the

---

curacy, or to take data from spatially separated points. Eventually, the experimentalist may decide that there is no low-dimensional attractor at all. I have benefitted from discussions with Jerry Gollub and James Crutchfield on this point.

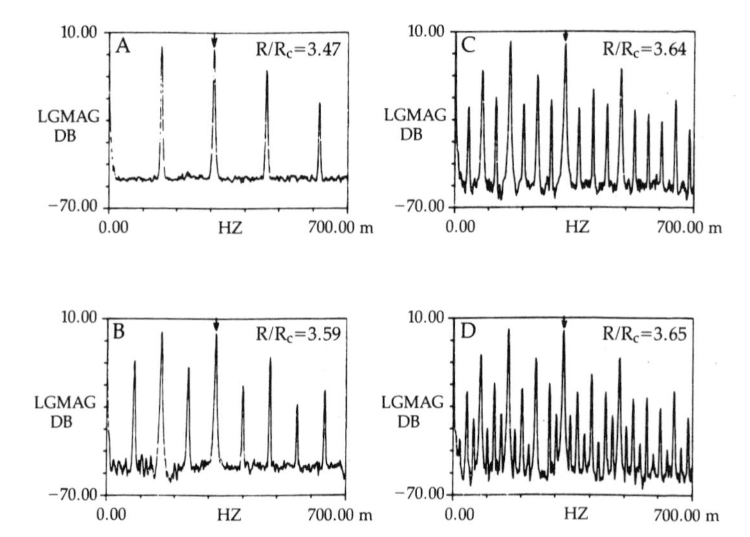

Fig. 5. Period-doubling in fluid convection. From Libchaber, Fauve, and Laroche 1983

position on the next pass through, and we do this repeatedly. "If the trajectory position were completely randomized each time around the attractor, one would expect a random scatter in such a plot. The fact that actually the points in such a plot fit along a well-defined curve" will show that the irregular behavior is deterministic chaos (Shaw 1981b, 224).

This method, which produces what is known as a "first-return map," effectively reduces the study of the system to an analysis of an iterated one-dimensional discrete mapping. If the first-return map has a quadratic extremum, for instance, the entire analysis of period-doubling can be applied (Bergé, Pomeau, and Vidal 1984, 219). Moreover, the first-return map can provide a measure of the Lyapunov exponent. By fitting a curve to the points on the map and then averaging the slope over the curve, an approximate measure of the degree of sensitive dependence on initial conditions can be obtained (Shaw 1981b, 224).

Figures 5 and 6 illustrate many of these experimental tech-

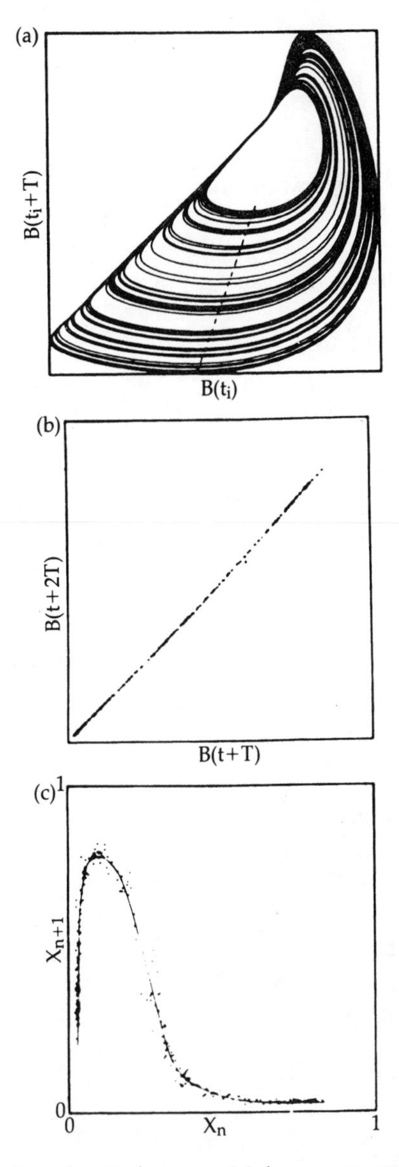

Fig. 6. Chaos in a chemical system. (a) A reconstructed attractor for the system. (b) A Poincaré surface-of-section through the dashed line in (a). (c) A first-return map. From Roux, Simoyi, and Swinney 1983

niques for analyzing data from physical systems. Figure 5 shows the Fourier power spectra of measurements of instantaneous velocity in a cell of fluid undergoing Rayleigh-Bénard convection. As the temperature gradient (represented by the parameter R/R_c) is increased, the fluid in the cell undergoes convection that displays period-doubling. The appearance of more and more "spikes" in the power spectrum demonstrates the onset of signals with frequencies $f/2$, $f/4$, $f/8$, etc. (Libchaber, Fauve, and Laroche 1983, 77).

Figure 6 depicts an attractor reconstructed from time-series data obtained by measurements of ion concentrations in an oscillating chemical reaction. It depicts a two-dimensional projection of the attractor, a surface-of-section view that appears as a line segment, and the first-return map that exhibits an extremum (Roux, Simoyi, and Swinney 1983, 262).

The three routes to chaos, sensitive dependence on initial conditions, strange attractors with elaborately folded fractal structure, and other elements of chaos theory have been reported in experimental systems. These systems vary from measurements of brain wave activity to yearly patterns of measles outbreaks to instabilities in the electrical conductivity of crystals to the wobbling of certain coffee-table toys (see Holden 1986; Hao 1984). Some of these examples of chaotic behavior have been convincingly documented in laboratory settings, while some of the examples of low-dimensional chaos outside the laboratory are still the subject of lively debate. But the interplay of theory and practice, aided by computer simulations, continues to expand the repertoire of models and techniques. "Chaos" proves to be a useful term indeed in a wide variety of scientific investigations.

2

Varieties of the Impossible

> Go ahead, ask yourself how and realize just how many factors had to be taken into account . . . so that finally the task would loom like a monster in your head, the reminder of . . . the limit of human intelligence and the fear of a curiosity which might at any time throw the world back into your face.
>
> —Jacques Menasché (1987)

The Bad News about Prediction

Some researchers see systems with sensitive dependence on initial conditions as important sources of novelty and diversity in the natural world. Chaotic phenomena guard our weather from falling into an ever-repeating rut and give rise to the variations underlying evolutionary change. But many physical scientists speak of chaos as another realm where human knowledge runs up against limitations. These limitations may be spoken of with fascination or awe, but we often discern a note of regretful humility, as Nature has decreed: "Here you can go no further." Christian Vidal writes that through the concept of sensitive dependence, "science is once again [in] the process of recognizing its own limits," for this idea "forbids us from ever being able to predict the destiny of a dynamical system whose flow is on a strange attractor, no matter what we do" (Bergé, Pomeau, and Vidal 1984, 268).

But precisely what kind of limitation on scientific knowledge does sensitive dependence impose? What exactly is forbidden and how does this restriction arise? To answer this, my first step will be to spell out just what the limitation asserts about predictability: chaotic systems require impossibly great resources for accomplishing useful predictions. Next I will clarify the meaning of this assertion—and especially of the words "useful" and "impossible." Finally, I will examine the views that originally led scientists not to expect such a limitation. Chief among these views is the methodological assumption that small errors in specifying initial conditions will remain small in future computations. By exploring what underlies this assumption, we can better understand the implications of chaos theory's challenge to it. The broad scope of the commitments underpinning this assumption will help explain why posing that challenge is often seen as the bearing of unfortunate news.

Chaos and Predictive Hopelessness

In seeking to clarify how and why predictability can be impossible, it will help to compare chaos theory with two other examples from physics in which certain kinds of knowledge are restricted or prohibited. In special relativity we find that we are utterly unable to assign an absolute temporal order to all events, and quantum mechanics (under some interpretations) demands that we refrain from attributing definite values to some properties of particles. These epistemic limits are due to certain scientific facts; because of the way the world is, no amount of effort will suffice to establish an absolute ordering of events in time, and no amount of investigation could suffice to determine both the position and momentum of a particle with arbitrary accuracy. Can the strictures of chaos theory be expressed in a similar way? It does not seem so, for to say that no amount of effort could suffice to predict the future state of a chaotic system is simply not accurate.

Chaos theory does not simply postulate the impossibility of prediction for certain systems, nor does it introduce any new laws of motion from which to deduce such a statement of limitation. In fact, chaos theory introduces no new postulates about the physical world at all. Nothing in chaos theory is analogous to the postulated finite speed of light in special relativity, or the postulates about measurement and noncommuting operators in quantum theory. Instead, chaotic models of physical systems are built in a strictly "classical" world using modeling equations such as the Navier-Stokes formulas for fluid flow. Such equations describe medium-sized objects traveling at moderate speeds in a thoroughly Newtonian manner. Yet from these systems and their characteristic sensitive dependence on initial conditions arises a limitation on our knowledge.

To see how this limitation arises, consider a dynamical system consisting of a set of variables $\{x_i\}$ that fix the state of the system in state space and a set of evolution equations that dictate how the system changes in time. Each of the variables x_i will be derived from the properties of the system being modeled: temperature, ion concentration, population density, etc. Such quantities are often real numbers, which means that to express them with our ten digits usually requires an infinite number of decimal places. Recall that in the world of classical physics the velocity of a particle may be represented as, for instance, $7.32890592 \pm 5 \times 10^{-8}$ m/s, but the actual value will almost certainly be irrational and thus require a nonrepeating decimal expansion.

So the values of the variables in our system can be fixed only within some finite degree of accuracy—we may measure the temperature as accurately as we like, but we will never succeed in writing down the infinite decimal representation of the real number involved.[1] Likewise, predictions garnered by

1. This constraint raises an important point about what is meant by "inaccuracy." In a physical science, inaccuracy can be due to equipment error, ineliminable perturbations from outside influences, or even the inherent indeterminacy of the quantity being measured. Above and beyond these is mathematical round-off error due to the fact that we simply cannot represent the full decimal expansion of a real number in most cases. All these sources of our inability to specify exactly the value of a variable fall under what I will call "vagueness."

working the evolution equations on a set of initial conditions can never be infinitely accurate. This constraint is nothing new; scientists routinely specify the margin of error in their measurements and predictions.

So in some sense we have never been able to make completely accurate predictions—where "completely accurate" would mean specifying the real-number value of each variable at some future time. Mathematical physical science has always been able to give only an approximate value and an assurance that the remaining decimal places could be found out with sufficient resources. So we cannot cast the limitation posed by chaotic systems as saying "completely accurate predictions are impossible," for physics never pretended to make completely accurate predictions. The predictions sought and found were always of the form: "given initial conditions with some specified degree of accuracy, determine the final conditions within some other specified degree of accuracy."

Given this realization, we can now define what is meant by a "prediction task" in order to clarify how sensitive dependence renders some such tasks impossible. For a dynamical system, a *prediction task* asks us to predict the value of a variable x at time t to within accuracy $\pm\delta$, given the values of the variables x_i at t_0 to within accuracies $\pm\delta_i$. This definition accords with what the philosopher of science Karl Popper calls the "principle of accountability," namely, that, given the degree of accuracy required in our prediction, we should be able to specify the degree of accuracy needed in our initial conditions (Popper 1956, 12).

The Lyapunov exponent of a system gives us a way of meeting just this requirement. For instance, in the Lorenz equations any vagueness in specifying one of the initial conditions will grow as $e^{\lambda t}$, where $\lambda \approx 1$. So, to know the state of the system 200 units of time into the future with an accuracy of ± 0.1, we must specify the current state with an accuracy of $0.1/e^{200}$ or roughly 1×10^{-88}. True, that is more accurate than any measurement ever made, but that does not suffice to say such a prediction is impossible.

Now of course this mode of calculating required accuracies is very abbreviated, for no mention is made of any factor other than the brute mathematical requirements of the Lyapunov exponent. I am neglecting, for instance, any mention of additional accuracy that may be necessitated by errors produced by the computational methods used. But this reasoning can construct a picture that forcefully argues for the impossibility of some prediction tasks. In doing so, I will use the term "predictive hopelessness" to describe a situation in which those predictions we can make fail to provide any information about the system whose behavior we are seeking to predict. As noted, any prediction has a certain degree of inaccuracy or vagueness associated with it. Imagine a situation in which our knowledge of initial conditions is only enough to give us predictions with an enormous spread of uncertainty or vagueness. If our best predictions are so inaccurate that the standard deviation of this "spread" is roughly the same size as the allowable state-space region, we face predictive hopelessness.

Recall that chaotic systems are bounded: they do not expand out to infinity but remain confined to the close neighborhood of some attractor (or energy surface, in the case of conservative systems). Each variable x_i will thus have some span Sp_i that it does not exceed. This fact allows us to define *predictive hopelessness* as follows: if any prediction task that asks the value of x_i with an accuracy greater than $Sp_i/2$ is impossible, then that complex of system-plus-initial-conditions is in a situation of predictive hopelessness with respect to variable x_i. For example, the Lorenz system has two lobes, one corresponding roughly to clockwise rotation of a fluid and the other to counterclockwise rotation. If the margin of error in my best prediction is so great that I cannot even tell which way the flow is going, then I am in a situation of predictive hopelessness. I may as well flip a coin.

Using this idea of predictive hopelessness, I will define the "predictively worthwhile time," T_w, for a dynamical system. When we know the initial values of the variables x_i each within a range of accuracy δ_i, the predictively worthwhile time tells us the time it takes for the situation to reach predictive hopelessness. $T_{w,i}$ is a function of the δ_i and tells us how long it will be before we must throw up our hands about the value of variable x_i. More decimal places of accuracy in our initial conditions will give us a longer predictively worthwhile time, while demands for more accurate predictions or any increase in the Lyapunov exponent will shorten the predictively worthwhile time. For some systems we may have data so inaccurate that the $T_{w,i}$ are smaller than the amount of time between when we measure the x_i and when we would want to predict their values. Thus, our measurements are worthless for prediction and the only way to escape this predictive hopelessness would be to make more accurate measurements.

For most systems in physics textbooks, our measurements can easily give us predictively worthwhile times of years—or even centuries, as in the case of predicting eclipses. In these sorts of situations, the predictively worthwhile time is much greater than the characteristic time between noteworthy events. With eclipses it is relatively easy to specify the time scale associated with important events, since they happen every few years. We have no straightforward way to associate a characteristic time scale with all systems in the natural world, but in general smaller systems have interesting things happening more often.[2]

But for some chaotic systems the sensitivity to initial conditions may be so great that in order to have a predictively worthwhile time greater than the characteristic time scale of the system's events, the initial conditions would have to be specified with an accuracy impossible to attain. Now moderately accurate initial conditions will always allow us to predict the system's behavior for some short time into the future, but

2. Characteristic time scales for subatomic interactions and chemical reactions are generally well agreed upon. For other physical systems there may be no standard answer to the question "how far into the future must we be able to predict in order for the prediction to be of any use." The implications of this term "useful" will be dealt with in the next section.

unless our predictively worthwhile time is greater than the system's characteristic time scale, our prediction is not a useful one; it does no good to be able to predict whether or not there will be an eclipse tomorrow if what we want is to be able to predict just when the next eclipse will occur. Since the vagueness in our initial conditions grows exponentially with time according to the Lyapunov exponent, we can see that for some chaotic systems with a long characteristic time scale, a useful prediction would require initial conditions to be specified with more accuracy than is possible. This, then, is the limitation posed by sensitive dependence: *chaotic systems require impossible accuracy for useful prediction tasks.*

This does not mean that all chaotic systems require impossible accuracy for all useful prediction tasks; some systems may manifest a mixture of predictable and unpredictable behaviors. But when we speak of chaotic phenomena as being impossible to predict, I would maintain that this is a good way to spell out that limitation.

Three Forms of Impossibility

Turning now to the question of what this limitation means, I will first confront the notion that it is a limitation that comes from us and is therefore not an "objective" limitation. This notion may arise from the use of two terms in the definition of this limitation: "useful" and "impossible." Both these words sound as if they should be followed by the phrase "to us," so that the formulation would read: "chaotic systems require an accuracy impossible to us in order to accomplish prediction tasks useful to us." Such a reading makes the limitation sound as if it affects us only because of a particular way we happen to be constructed—it is an epistemic limitation on the knowledge we can gain given certain peculiarities about what is useful or attainable for us.

Let us address first those qualms that may be raised by the word "useful." It may indeed seem that since the limitation is expressed in terms of disallowing some "useful" predictions, it is not a limitation built into the structure of the world but rather a limitation that exists solely because of some interests we happen to have. Such a limitation may seem to be an artifact of the way we are and thus different from the more absolute-sounding limitation posed by, say, special relativity. In the case of chaos theory, the boundary between "useful" and "not useful" predictions does not instantly appear before our eyes—it is a line that must be drawn by us, based on the kinds of things we want to know. But once we have drawn that line, for instance by deciding the characteristic time scale for important events in a system, a simple calculation using the Lyapunov exponent will tell us whether there is any hope of making an informative prediction. The limitation appears with all the mathematical, impersonal inexorability needed to satisfy us that it is objective. The appearance of the word "useful" in the above formulation, a word whose demarcation requires explicit mention of human interests, may thus render the statement of the limitation "human-dependent." But the limitation is not "human-dependent" in any subjective or merely conventional sense. It is not up to us.

The question of "human-dependence" goes even further than concern over the word "useful," for it may well seem that the limitation posed by chaotic systems is only a limitation for us due to our limited resources. After all, a degree of accuracy impossible for us to attain may well be within the reach of scientists in the future. The concern here is over the meaning of the word "impossible"; what exactly is the nature of the impossibly great accuracy needed to make useful predictions of a chaotic system? Is it that such accuracy is simply beyond the reach of our current machinery, or is there some fact about the physical world that renders such accuracy forever beyond our reach?

To help clarify the meaning of this impossibility, I shall first distinguish three types of situations in which we are unable to do something. First, *logical* impossibility will apply to those tasks where success would involve a manifest logical contradiction: it is impossible to make a rock that is both black and not black in the same way at the same time. Such a feat is impossible throughout our world and, some would say, in any world we can imagine. Next, *theoretical* impossibility would apply to those tasks where success would violate an accepted law of nature: it is impossible to build a perpetual motion machine or travel faster than the speed of light. Since physical laws are held to be valid in all times and places, a theoretically impossible feat will be impossible throughout the world, independent of human capabilities. Finally, *practical* impossibility would apply to tasks whose completion would require more resources than are available to human beings: it is impossible to construct another earth-sized planet to relieve global overcrowding. We cannot accomplish such a feat now and there is no way we could accomplish it in the near future. But this impossibility does not hold for all times and places; humans in another time (or other intelligent creatures) may be able to do such a thing. So this statement of impossibility is relative to a certain time and place and configuration of human abilities.

Having established these categories, it may seem straightforward to classify the limitation posed by chaotic systems as a practical impossibility. After all, we have already noted that chaos theory introduces no changes in the laws of physics, so there is no new principle that allows us to say that useful predictions are impossible "in principle" or on theoretical grounds. Yet at least one physicist has argued that our inability to predict chaotic behavior stems from more than just technological insufficiency. "The necessity of recourse to statistical methods" is here "no longer justified merely by considerations of a pragmatic nature, but is imposed, as it were, by mathematical logic itself" (Bergé, Pomeau, and Vidal 1984, 268).

In what follows I shall argue that the limitation under discussion is indeed as much a matter of theoretical as practical impossibility. In fact, this limitation straddles the distinction as presented and impels us to reconsider this initially plausible formulation of types of impossibility.[3] To see this, consider a typical chaotic system such as turbulent convection in a fluid. If I say that, given the state of the system now, we cannot predict its state in one minute, this may be a clear case of a practical impossibility. After all, it is conceivable that with hyperaccurate velocity probes it will one day be possible to make such a prediction correctly. But once we realize that "predicting a later state of the system with a specified accuracy will require that uncertainty of initial conditions decrease as an exponential function of the timespan of prediction," it becomes easy to imagine prediction tasks that would require mind-boggling accuracy (Thompson and Stewart 1986, 225). For example, consider the earth's atmosphere, a large system with turbulent fluid flow. If I ask a question far enough in the future (e.g., "where on the Atlantic coast will the first hurricane of the year 2093 strike?"), a successful prediction could require astronomical accuracy—perhaps more bits of information specifying initial conditions than there are electrons in our entire galaxy. Now, is our inability to achieve this accuracy a *practical* limitation?

At what point does the number of accurate decimal places required for predictions become so great that our inability to achieve such precision is no longer just a matter of our limited resources, but a scientific fact about the way things are? Some may contend that so long as the prediction task for a chaotic system requires only some finite degree of accuracy, it is incorrect to say that such accuracy is impossible to achieve. (Recall that we are dealing strictly within the world of classical dynamics here, where there is no barrier to claiming that the

3. The blurring of the practical/theoretical distinction under discussion here is not a challenge to the traditional division between practical and theoretical *activity*. Chaos theory is very much a theory in the sense of abstracting, universalizing inquiry. But to the extent that practical impossibilities are viewed as one kind of problem while questions of theoretical possibility are part of another field of study, the discussion in this section may be relevant to the broader issue.

values of physical quantities are determinate real numbers.) After all, we can imagine some highly advanced human or alien intelligences who could measure and calculate with accuracies far beyond our own abilities. If such beings could accomplish all prediction tasks for chaotic systems, then our inability to do so is only a practical limitation after all.

In response to such a position, I shall present a series of arguments that this limitation is not simply a matter of practical lack of resources. First we should agree that it is not enough to be able to imagine that someone or something could accomplish any given prediction task. That we can imagine it merely shows that such a feat is not logically impossible. After all, we can imagine someone traveling faster than the speed of light or measuring the position and momentum of a particle simultaneously and with great accuracy. The question is whether we can imagine someone accomplishing such a feat without violating any laws of physics. If the contention is made that predictions about a chaotic system are imaginable and that our inability to achieve them is thus due only to a lack of resources or ingenuity on our part, then a crucial step is missing. For if it can be shown that such a prediction would involve a physical impossibility, the ability to imagine success will not mean that our failure is due to practical constraints.

Next let us consider a very prosaic way in which success at some prediction tasks could violate physical law. Let us say there are N electrons in our galaxy. It should not be hard to construct a prediction task for a chaotic system that asked for the state of the system so far in the future that the number of binary digits needed to specify the initial conditions would exceed N. This kind of reasoning can be followed in a number of directions. If we use other galaxies to store and compute additional digits, it may take too long for signals to travel around our giant computer, and the predictions may be unavailable until the event that was to be predicted already took place. If the galaxies are moved closer together, gravitational collapse may wreck the machinery. The point is that a task that grows exponentially in difficulty can always be made so hard that the universe simply cannot provide enough mass or energy or even room to write down the numbers involved.

After all, the prediction tasks we are speaking of are tasks for beings with finite abilities. A finite being can accomplish only finite tasks. While it may be true that for any prediction task, some imaginable superpowerful-yet-finite beings could accomplish it, it is also true that for any finitely powerful being there is some specifiable prediction task that it cannot perform. The exponential growth in the difficulty of prediction tasks as the time-frame or Lyapunov exponent increases makes it relatively simple to go one better than any finite range of capabilities specified. This ploy of inflating the requirements for accuracy can be kept up until one is simply forced to admit that there is more involved than just a lack of technology or cunning. Some prediction tasks just cannot be accomplished because to do so would require resources that are physically impossible to obtain.[4]

And if some interlocutor would refuse to submit in the face of astronomical requirements for accuracy, implacably insisting on imagining trillions of cooperative neutrino-beings wielding computers the size of the known universe, then and only then does the trump card of quantum-mechanical uncertainty need to be played.[5] But before that happens, it should

4. Another example of a task that requires impossible resources could be time travel into the spatially nearby past. David Malament has demonstrated that although such trips do not violate the physical laws of a universe with Gödelian space-time structure, they would require a dizzying amount of fuel (Malament 1985). So much fuel, in fact, that Paul Horwich writes that such trips would "always be technologically impossible" (Horwich 1987, 122). I submit that a task which would *always* be technologically impossible is well on the way to theoretical impossibility.

5. Ironically, once quantum-mechanical uncertainty is brought into this discussion, the game is over. For if an arbitrarily accurate statement of initial conditions is ruled impossible because of quantum-mechanical effects, then our inability to predict the future state of the system becomes a clear-cut case of theoretical impossibility.

be clear that what makes some prediction tasks impossible is some fact about us—in the sense of finite beings in this physical universe—and not *merely* some fact about us—in the sense of just us poor humans with our current historically limited resources. Some systems are predictable in principle because they evolve according to strictly deterministic equations that render their behavior calculable in the very short term, and yet these same systems are unpredictable in principle because our inability to make such calculations in the longer term is not due to some constraint we could overcome.

One possible implication of this kind of limitation is that we may be pushed into indeterminism, for if these predictions are impossible in principle, we can no longer accept that the random or chancy nature of chaotic systems is merely an artifact of our human failings. It becomes much harder to hold that there is some already-set sequence of events that we cannot see only because our senses and our instruments are not precise enough; it no longer seems right to label the patternless behavior of the weather "*apparent* randomness" as if we could in fact predict the weather perfectly if only we had enough data. This line of reasoning, which may seem verificationist, concludes by stating that if the pattern is in principle impossible to predict, then there is no underlying pattern there at all. Whether this is the case or in-principle unpredictability and determinism can in some sense cohabit will be dealt with in chapter 3.

Before considering that question, one more clarification needs to be made. The earlier distinction between theoretical and practical impossibility must be revised. Chaos theory discloses a region of logical possibility closed to us neither by physical law nor by limited resources, but by the fact that we are finite beings. Our finitude does not derive directly from any law or postulate of physical science, but neither is it a merely practical limitation that we could someday shed. It is an important fact about the way our intelligence functions in this world, and chaos theory is interesting because it makes our finitude have real consequences for our science. So the "recourse to statistical methods" is not, strictly speaking, forced upon us "by mathematical logic itself"; it is forced upon us by the conjunction of mathematics and this fact about us—the fact that we cannot be as accurate as we want.

Seeing the limitation in this way may provide a reason to reassess the three types of impossibility. We may wish to introduce a fourth category of impossibility to account for the distinctive type of limitation we find in chaos theory (and perhaps elsewhere). I propose that we borrow from Kant and label this type of impossibility "transcendental."[6] *Transcendental* impossibility would apply to those tasks where success would be inconsistent with nonaccidental facts about human inquiry; it is transcendentally impossible to accomplish some long-term prediction tasks for chaotic systems. Transcendental impossibility holds for all configurations of human abilities that retain such nonaccidental features of human inquiry as the following: being conducted by finite physical organisms, being expressed in language, and being motivated by interests and values.

But perhaps redefining and adjusting the categories of impossibility is not quite so enlightening as questioning the basis upon which the categories were constructed. Recall that the three types of impossibility rested on a separation of the logical structure of the world from the physical structure of the world and yet another separation between the structures of the world and the structures of us. Perhaps there is little to gain from seeking to place the blame for every limitation of our inquiry on one or the other, the world or us. Chaos theory suggests that limitations come from just that complex of the world and us within which all inquiry takes place.

6. "Transcendental" here does not mean "passing beyond all possible experience," but "preceding all experience as a precondition for knowledge" (paraphrasing Kant [1783] 1950, 123).

Between Methodology and Metaphysics

The preceding discussion has clarified the nature of the limitation posed by sensitive dependence, but a question remains: Why was this limitation so unexpected and so unwelcome? In chapter 5 I will consider the historical question of why chaotic behavior took so long to be studied, but here I will argue that the limitation posed by chaotic systems presents a troublesome challenge to a certain methodological assumption that has long guided scientific practice. This assumption, that a small amount of vagueness in measurements will lead to only a small amount of vagueness in predictions, meets a direct challenge from systems with sensitive dependence on initial conditions. As Arthur Winfree, a theoretical biologist, has stated:

> The basic idea of Western science is that you don't have to take into account the falling of a leaf on some planet in another galaxy when you're trying to account for the motion of a billiard ball on a pool table on earth. Very small influences can be neglected. There's a convergence in the way things work, and arbitrarily small influences don't blow up to have arbitrarily large effects. (Gleick 1987, 15)

Of course, the "blowing up" of arbitrarily small influences is precisely what happens in systems with sensitive dependence.

In this way, chaos theory can be seen as presenting a further limitation of the scope of classical physics: Newtonian mechanics breaks down at high speeds, small sizes, and now at the transition to chaos as well. But the boundary of the applicability of classical physics does not have to be redrawn because of the failure of some fundamental laws. For it is not that the equations of motion must now be written in new ways or that basic concepts like mass or energy need to be redefined. Recall that a premier chaotic system like that of Lorenz is based upon a set of equations describing scrupulously Newtonian behavior.

Neither is it the case that chaotic systems force a "breakdown" in the applicability of classical physics because of their horribly complicated mathematics. For in the first place, many nonchaotic systems lack a straightforward closed-form solution to their evolution equations. And in the second place, even if we could obtain a neat and tidy solution to the equations in a chaotic dynamical system, such a solution would be effectively worthless since reliable predictions would require unattainable accuracy. This limitation is the special character of chaotic systems that causes so much distress. Usually when we are confronted with an insoluble set of equations, we can approximate to some similar system whose behavior we can predict. This elaborately developed approximative technique goes by the name of "perturbation analysis," and it treats a difficult case as a version of a well-understood situation plus some small "perturbation" that complicates the situation but keeps it mathematically tractable (see, for example, Marion 1970, 165–71). In making this move, we trust that slight discrepancies and higher-order terms will not render our approximation worthless. But sensitive dependence guarantees that such approximations will be worthless. Thus, we can characterize the challenge to classical mechanics in this way: *As chaos sets in, we encounter the inadequacy of our methods, not the inadequacy of our laws.*

The limitation posed by chaos theory now presents itself as a challenge to a methodological assumption. This assumption, codified by the French mathematician Jacques Hadamard, is a stability condition on the solutions of systems: in a well-formulated mathematical representation of a natural system, small differences in initial conditions will not grow into large differences in later trajectories; if they do, then the system does not accurately describe the natural system it models (Earman 1986, 154; Hadamard 1952). In other words, sensitive dependence is seen as an aberration that signaled to the researcher that the mathematical model was defective. This belief is methodological in character because it serves explicitly to guide scientists in the posing of questions and the search for answers.

To understand better why chaos theory requires a revision in methodology, we should draw a clear distinction between method, methodology, and epistemology. A *method* is a procedure for collecting evidence, a *methodology* is a theoretical analysis of the research process (which can be descriptive or normative), and an *epistemology* is a theory of the nature and scope of knowledge.[7] But the assumption described above purports to say something quite general about the content and structure of the world—as such it seems to be a metaphysical assumption. Consequently, we will need to clarify what is meant by asserting that chaos theory issues a methodological challenge and not a metaphysical one.

I will begin by exhibiting some clearly metaphysical tenets. These will include very general statements about what exists—statements that constitute an ontology. Some examples are "minds exist," "neutrinos exist," or "whatever entities make our science most beautiful exist." Then there are very general statements about the structure of the world, which make up another part of metaphysics—cosmology. Examples of these include "things change their states in discrete jumps"; "there are no uncaused events"; and "time is absolute." By this account, the beliefs that carry the labels "dualism" and "materialism" are ontological; beliefs in determinism and spatial continuity are cosmological.[8]

What then are methodological beliefs? They are very general statements about how to understand the world; for example, "physical quantities can be represented successfully by real numbers" and "physical change can be understood using

7. See Harding 1987, 2–3.

8. This formulation of the meaning of metaphysics is clearly impoverished in comparison with the view that metaphysics seeks the meaning of Being. But chaos theory as a current field of scientific research can shed light on metaphysics conceived as only a set of very general statements about beings.

differential equations." Now tenets such as these are not easily separable from metaphysical beliefs—the former statement is connected to a cosmological belief in continuity and an ontological commitment to measurable properties, the latter is connected to a cosmological belief in determinism.

The methodological assumption undermined by chaos theory—that a good representation of a system will not display sensitive dependence on initial conditions—is likewise connected to metaphysical beliefs. This connection is not one of straightforward logical dependence, such that the success of chaos theory can disprove some metaphysical tenet. Instead, chaos theory plays a role in an interdependent system of beliefs. But methodological beliefs are explicitly normative; while metaphysics limits our perceptions and guides our practices "behind the scenes," methodological tenets are more up front about it. Our answers to questions such as "what is there" and "how does it act" are more likely to remain unspoken than are our answers to the question "what should we do to find out?" Metaphysics is abstracted from actual practices, while methodology involves the concretization of some metaphysical beliefs as principles guiding our research. Such principles assert that it is possible or advisable to do something, and this assertion of course rests on a belief that the world contains certain things or behaves in certain ways. Thus, methodological tenets based on these metaphysical ideas can be useful or not, and if they are no longer useful this can lead us to reconsider the applicability or appropriateness of the various metaphysical beliefs undergirding them.

Some of the beliefs that lie behind the methodological trust in the stability of small amounts of vagueness are these: first, a belief that we can understand systems in isolation—that all elements of the world are not so interconnected that any attempt to ignore "outside" influences is doomed. This principle of isolation is related to the methodological tendency that the philosopher Paul Teller has termed "particularism"—the view that the universe should be approached as a collection of individual entities with nonrelational properties

(Teller 1986). Next is a belief that simplicity in mathematical representation goes hand in hand with simplicity of behavior—the view that once we can capture the behavior of a system with a simple rule we have understood that system. This belief draws support from and reinforces a view of understanding as identical with exact prediction. The connections between chaos theory and these beliefs will be further explored in chapters 4 and 5.

Finally, there are the two beliefs that can be called "determinateness" and "determinism." Determinateness is the belief that physical properties are definite and determinate—that properties do indeed have a pre-existing and infinite degree of accuracy, which justifies their representation by real numbers. And determinism is the belief that the universe changes in time like the unique solution to one grand system of differential equations. The relation between chaos theory and these last two methodological and metaphysical doctrines will be the subject of the next chapter.

The Challenge of Openness

Chaos presents a limitation on the predictability of physical systems because sensitive dependence creates a requirement of impossible accuracy in order to perform useful prediction tasks. This limitation straddles the putative line between theoretical and practical impossibility by presenting us with examples of tasks so difficult that the very fact that we are finite beings makes us unable to accomplish them. Chaos theory thus challenges the very distinction between theoretical and practical impossibility; our inability to perform certain prediction tasks is due neither to the strictures of physical law nor to mere technical difficulty. And chaos theory provides not only a limitation on predictability but a new limitation on the applicability of classical physics. Sensitive dependence on initial conditions undermines the justification for the methodological assumption behind the standard strategies of approximation and perturbation analysis. Impugning this methodological assumption calls into question a number of beliefs about the way the world is and how to go about understanding it. For if our world contains chaos, even the smallest vagueness can eventually blossom into utterly open ambiguity. The challenge of chaos may be to welcome this openness and not see it as cause for regret.

3

Unpeeling the Layers of Determinism

> And if you ask me how, wherefore, for what reason? I will answer you: Why, by chance! By the merest chance, as things do happen, lucky and unlucky, terrible or tender, important or unimportant; and even things which are neither, things so completely neutral in character that you would wonder why they do happen at all if you didn't know that they, too, carry in their insignificance the seeds of further incalculable chances.
>
> —Joseph Conrad (1914)

A Serious Revision

Christian Vidal, in *Order within Chaos,* writes that one element of chaos theory that is "rich in epistemological consequences" is "the challenge to the meaning and to the scope of the ideas of determinism and chance, as we are accustomed to practicing them today." "Obviously," he says, "a serious revision of their scientific definitions is imperative, and we must now go beyond the ideas stated precisely for the first time by Laplace close to two centuries ago" (Bergé, Pomeau, and Vidal 1984, 268). One important first step in revising these ideas has been to see how chaos theory drives a wedge between determinism and predictability.

While we once thought that a deterministic system and a predictable system were two names for the same thing, philosophers of science such as Mark Stone (1989), John Earman (1986), and G. M. K. Hunt (1987) have shown that chaos theory presents us with deterministic models of physical systems that are nonetheless unpredictable. By revising the concept of determinism so that it no longer must include the idea of predictability, these writers have sought to defend the doctrine of determinism, or at least keep it viable. In this chapter I will take up this issue and explore the varied senses of determinism at work in these discussions, with the goal of pushing the challenge to determinism even further. My belief is that chaos theory not only argues against the predictability of certain systems, but that when combined with quantum-mechanical considerations it leads us to grave doubts about the doctrine of determinism itself.

My strategy will be first to organize the many senses of determinism that operate in scientific and philosophical discussions about chaos theory. I will isolate four uses of the term and arrange them in a rough order, from the simpler and less restrictive to the more robust and full-blown. The question of whether determinism holds can operate at any of the following levels, or some combination of them:

A. Differential Dynamics. Are differential equations sufficient for description of the system?

B. Unique Evolution. Is the evolution of the system uniquely fixed once we specify the state of the system at any one moment?

C. Value Determinateness. Do all properties of the system have well-defined real values?

D. Total Predictability. Is the system predictable by a superior intelligence?

Next I will describe the philosophical argument that chaos theory encourages us to separate total predictability from our notion of determinism and how the special properties of chaotic systems force us to admit that not all aspects of the world are predictable. This discussion will be followed by a brief account of the role quantum mechanics plays in defeating the notion of value determinateness, further reducing the scope of the determinism at work in our world. Finally, I will demonstrate that chaos theory when considered along with quantum mechanics calls into question even the idea that our universe has a unique course of evolution.

Why Determinism?

Before entering into these considerations, it is worthwhile to ask, What motivates people to worry about determinism? and what important difference does it make whether determinism is valid or not? Addressing these questions pays proper respect to the injunction of Gaston Bachelard, the French philosopher of science, who wrote that ideas such as determinism "are properly studied in all the complexity of their psychological context, surrounded by all the empirical and emotional ambiguity that normally accompanies research on the frontiers of science" (Bachelard 1984, 99). Several issues generate philosophical interest in determinism, including the dream of total predictability, the denial of time and change, and the need for ultimate explanations.[1]

Starting with the first of these, we must acknowledge that many people's interest in determinism is bound up with the ideal of total predictability. True, the idea that determinism and predictability are identical is mistaken—at least some cogent senses of determinism do not imply predictability, and some nontrivial predictability does not imply determinism. But the two notions are nonetheless connected in an important way; determinism in the guise of differential dynamics, unique evolution, and value determinateness is a necessary, though not a sufficient condition for total predictability. If determinism is false—if this same "present" universe *could* evolve into more than one physically allowable "future" state—then we lose some predictability. So although the two notions are not

1. I have not included the determinism–free will controversy here, even though it is obviously a prime motivator of philosophical interest in determinism. It has always seemed to me that the freedom of the will counterposed to determinism is a kind of metaphysical freedom, and metaphysical freedom is not what there is a shortage of in the world today. So I will leave aside for now the issue of the implications of chaos theory for free will.

interchangeable, from a motivational point of view they are nonetheless linked, insofar as a challenge to determinism creates a challenge to the ideal of total predictability.

This goal of total predictability is neither all-important nor totally irrelevant. I do not mean to suggest that there is a single scientist who actually thinks *we* could ever construct a Laplacian supercomputer that would achieve the goal of omniscience. So this ideal does not operate as a concrete goal anyone actually hopes to achieve.

But neither is it merely a pipe dream devoid of motivational force. On the contrary, I think this goal represents a very real and precious ideal to many scientists, not something that might be reached but something to be approximated as closely as possible. For, as Nobel Prize-winning chemist Ilya Prigogine writes, total predictability was never "more than a theoretical possibility. Yet in some sense this unlimited predictability was an essential element of the scientific picture of the physical world. We may perhaps even call it the founding myth of classical science" (Prigogine 1980, 214).[2]

In chapter 4, I will discuss the problematic identification of understanding with prediction. For now, it should be clear that if understanding means being able to predict, then the (admirable) goal of increasing understanding may also be seen (by practitioners) as a quest for getting closer to total understanding—total predictability. Any threat to the very possibility or theoretical coherence of that goal can be seen as a threat to the "aim of science." But that should happen only for those who think *improving* something means heading toward some in-principle *final* state. I do not.[3]

2. In chap. 5 I will discuss the connection between a scientific interest in predictability and a social interest in the control of natural processes. I will argue that the "founding myth" of total predictability exerted an ideological influence that contributed to the delayed development of chaos theory. For another discussion of the connection between determinism and the ideal of omniscience, see Husserl 1970, 65.

3. Thomas Kuhn argues against the necessity of positing an ultimate goal of a completed science, and he maintains that we can make sense of science

The next need that motivates discussions of determinism is the impulse to collapse past, present, and future into one co-present continuum—the need to do away with time. In his book on determinism, *The Open Universe,* Karl Popper locates the source of this notion in theological doctrines of divine omniscience and gives an excellent portrayal of the resulting deterministic worldview:

> The intuitive idea of determinism may be summed up by saying that the world is like a motion-picture film: the picture or still which is just being projected is *the present.* Those parts of the film which have already been shown constitute *the past.* And those which have not yet been shown constitute *the future.*
>
> In the film, the future co-exists with the past; and the future is fixed, in exactly the same sense as the past. Though the spectator may not know the future, every future event, without exception, might in principle be known with certainty, exactly like the past, since it exists in the same sense in which the past exists. In fact, the future will be known to the producer of the film—to the Creator of the world. (Popper 1956, 5)

Perhaps, as Bachelard suggests, this view originated in the idea that the welter of mundane affairs is actually controlled and fated by the timeless workings of the heavenly realm, where "the feeling of *determination* was a feeling that a fundamental order exists, a feeling of intellectual repose stemming from the symmetries and certainties inherent in the mathematical analysis" (Bachelard 1984, 102).

The goal of ultimately successful scientific knowledge, for some, is to do away with our felt distinction that the present is "special" and different in kind from the past and future, and thereby come as close as possible to the true "God's-eye view" of the universe: "Once Einstein said that the problem

---

getting better, progressing, without recourse to some hypothetical picture of the end of inquiry (Kuhn 1962, chap. 12).

of the Now worried him seriously. He explained that the experience of the Now means something special for man, something essentially different from the past and the future, but that this important difference does not and cannot occur within physics" (Rudolf Carnap, quoted in Prigogine and Stengers 1984, 214). Determinism as a doctrine binds both past and future irremediably to the given present. All is given in one moment—the changing "now" is just our subjective window for experiencing the eternally present one instant at a time. For those who see any openness into possible futures as fatally threatening to the orderliness and intelligibility of the universe, this view is powerfully comforting. But it is also deeply disturbing because it does so much violence to our lived experience of time, change, and, yes, free will. The ability of determinism to provide a meeting place and arena of conflict for these competing needs puts it in a position of philosophical importance.

Finally, we can see determinism providing comfort to those who feel a need for ultimate explanations. The process of seeking explanations always seems to allow someone to ask "but why is *that* the case?" in response to any offered explanation of an event. Determinism surveys all the infinite series of such questions and assures us that there is indeed an ultimate reason.[4] After all, if things could not be otherwise, then the current total state of the universe, together with a full set of physical laws, comprise an adequate (though often overlong) explanation for each and every event. Of course this trades on a certain crypto-positivist definition of "explanation." Again, the question of understanding enters. If explaining an event means predicting (or retrodicting) it, then

4. Here I am avoiding a formulation of this motivation in terms of sufficiency or necessity. Any actual explanation will finally rely on "unexplained explainers" (see Toulmin 1961), and this is akin to a formulation of determinism that begins from a given set of "laws of nature." An even broader notion of determinism, similar to that which Earman discusses under the name "fate," would require that everything—including the laws of nature—*has to be* the way it is; see Earman 1986, 18–19.

determinism assures us that every event could be derived from one set of initial conditions plus laws. Without this, we confront a very real fear that some events may "just happen"—without any ultimate explanation, without any final reason, and seemingly without meaning. As John Earman writes, "if the only alternatives to determinism are final causes (e.g., divine intervention) and hazard (e.g., accident or chance), then determinism is attractive as an *a priori* truth or a methodological imperative of scientific inquiry" (Earman 1986, 23).

It is here that theological concerns enter, but I am afraid that I am not equipped to do justice to this issue. I believe that the opposing views of Monod (1971) and Bartholomew (1984) treat this problem—the problem of how an omnipotent God can be conceived of in a world with real chance—quite well.

The Four Layers of Determinism

When concepts akin to randomness and fate threaten to align or contradict (or even cohabit, in the form of terms like "deterministic chaos"), it will help to systematize some of the many senses of determinism used in discussions of chaos theory. My purpose is not to propose a single definition, but rather to enumerate the possible issues involved in the question, What does chaos theory tell us about determinism? The term "determinism" gets used in two ways: one describes a certain type of system and the other refers to a philosophical tenet of the type I call cosmological (as in chap. 2). The two uses are related: if a person believes that the whole universe, as a system, is deterministic, then that person subscribes to the tenet of determinism.[5]

The various senses of determinism can be arranged under the fourfold division described earlier, a scheme suggested by

5. The very notion that the universe as a whole can be treated as a "system" runs into philosophical difficulties familiar since Kant. However, I will not take up the issue of the additional assumptions necessary to move from "determinism is universally applicable to all systems" to "determinism is applicable to the universe as a system."

the work of Mark Stone (1989). This scheme allows us to isolate four of the key layers or levels of the complex concept of determinism and to see how this concept is built up from combinations of these layers. Each level can be applied to an isolated system or used to characterize the physical universe as a metaphysical tenet. Some writers, like Popper, hold that determinism really means the sum of all four layers. Other philosophers of science, such as Clark Glymour (1971) and Mark Stone (1989), limit the sense of determinism to differential dynamics, unique evolution, and value determinateness. Earman insists that only the first two are needed to capture the meaning of determinism, and I believe that differential dynamics by itself has much to recommend it as the best definition.[6]

Differential Dynamics

The first and simplest layer of determinism is the requirement that in a deterministic system the future depends on the present in a mathematically specifiable way. This layer may be thought of as growing out of the everyday sense of what it means to be "determined": the way a system is at this moment determines, establishes, and specifies the way it will be at the next moment. By this criterion, a system is deterministic if the way it changes in time can be specified by a set of differential equations.[7] Note that this is the very heart of what it means to be a (differentiable) dynamical system, and chaos theory is very much the study of dynamical systems. In this sense, the systems chaos theory studies are deterministic by definition (see Bergé, Pomeau, and Vidal 1984, 101; Earman 1986, 157).[8]

6. By "the best definition" I mean something like "the strongest characterization that is defensible in the context of modern physics." Thanks to Peter Kosso for suggesting this formulation.

7. Or a set of difference equations, for a system in discrete time.

8. Another sense of determinism is used by Bergé, Pomeau, and Vidal to mean "of a small number of degrees of freedom" or "having a low-

We can refine the sense in which differential dynamics means determinism by noting that a system is deterministic if the dynamical system that models it makes no reference to chance. Recall from chapter 1 that a dynamical system has two parts: a representation of all possible states of the system and a set of equations that describes how the state of the system changes with time. When neither of these two parts involves chance explicitly, we have a deterministic dynamical system.

For the state description, the first element of a dynamical system, to make no reference to chance, it must not be viewed as an average over many subsystems—we must be on the microlevel, not the macrolevel as in thermodynamics.[9] A state, by this account, must include all physical information about the system at a particular time; there must be no probabilities built into the state description because of an averaging over many indistinguishable substates. Any complex behavior in the dynamical system must come from its internal mathematical structure, not from the fact that the system is an approximation of a huge number of complicated interacting subsystems. The deterministic approach chaos theory takes to the study of apparently disorderly behavior is thus in contrast to a statistical approach, which focuses on the evolution of average values at many places in the system, or averages over many systems.

And for the second part of a dynamical system to make no reference to intrinsic probabilities, there must be no places

dimensional attractor." While this version does not seem easily translatable into a global or cosmological tenet, it has the advantage of specifying what some scientists are looking for: a way to describe very complex motion in simple terms. A system is thus seen as deterministic if some aspects of its evolution—say, its qualitative or topological character—are governed by strictly deterministic differential equations, yielding a kind of predictability-of-higher-order-characteristics; see chap. 4.

9. This is not meant to exclude systems such as hydrodynamic equations, which can be derived from averaging. Such equations are treated as macroscopic descriptions of one system.

where the instructions for finding the state at some future time say "flip a coin" or "40% go to state S_a and 60% go to state S_b." The rules or evolution equations must be deterministic in that they allow no branching, no choices, no plurality of possibilities; they must not be stochastic equations (Thompson and Stewart 1986, 188; Suppes 1984, 21). The requirements of differential dynamics satisfy this sense of determinism by insisting that the differential equations involved make no explicit reference to transition probabilities.[10]

These requirements make clear what differential dynamics means when applied to an individual system, but what would it mean when applied as a general principle? On the broadest scale, one could say that determinism as differential dynamics is just the tenet that we should keep looking for reasons for events in their pasts. The injunction is: use mathematical expressions (differential equations) to model the changes of physical systems; seek to understand or predict the future by relating it to the past with mathematical rules. Perhaps these rules will provide strictly unique implications and perhaps not, but keep trying to explain or predict until you cannot any more. According to this formulation, determinism is a methodological tenet based on a cosmological belief in the notion that the next moment in the evolution of the universe flows out of this one according to intelligible rules.

This tenet is difficult to spell out in a precise way, for it is not enough to say "The state of the universe at time t is related to the state at time $t_0 < t$ by a mathematical expression with differential equations," because any two quantitative formulations of the "state of the universe" are surely related by some function or other. To put teeth into this notion, it may be necessary to specify that the mathematical rule must involve

10. This second requirement can also be formulated as requiring the evolution equations to map pure states onto pure states. There must never be a situation where a complete specification of the state at one time will evolve into a mixed state characterized by a probabilistic distribution of properties. The second requirement thus ensures that if the first requirement is met at some time, it will not be "undone" by the working of the evolution equations.

simple, time-invariant differential equations. I will return to the question of the meaning of this layer of determinism after considering some of its companions.

Unique Evolution

John Earman calls the next level of determinism "Laplacian determinism," and it consists of the uniqueness of the universe's timeline: if two worlds agree on everything at one time, they must agree on everything at all other times.

> Letting $\mathcal{W}$ stand for the collection of all physically possible worlds, that is, possible worlds which satisfy the natural laws obtaining in the actual world, we can define the Laplacian variety of determinism as follows. The world $W \in \mathcal{W}$ is Laplacian deterministic just in case for any $W' \in \mathcal{W}$, if W and W' agree at any time, then they agree for all times. By assumption, the world-at-a-given-time is an invariantly meaningful notion and agreement of worlds at a time means agreement at that time on all relevant physical properties. (Earman 1986, 13)[11]

This layer of determinism holds that the complete instantaneous description of a deterministic system "fixes" the past and future with no alternatives. On the level of a metaphysical tenet, this simply means that the universe as a whole is a deterministic system.[12]

11. Note that dissipative systems will defeat this kind of determinism unless there is "room" at infinitesimal scales in state space for all trajectories to fit into. In classical dynamics, each set of initial conditions will follow a unique trajectory, but in an isolated dissipative system, "*all* non-equilibrium situations produce evolution toward the *same* kind of equilibrium state. By the time equilibrium has been reached, the system has *forgotten* its initial conditions—that is, the way it had been prepared" (Prigogine and Stengers 1984, 121). Similarly, for dissipative systems the flow "is not invertible because of the loss of historical determinism" (Earman 1986, 164).

12. This layer of determinism may be best considered on the scale of a metaphysical tenet applying to the entire universe because, as John Ear-

Value Determinateness

Numerous writers regard the uniqueness of a system's evolution as only part of the meaning of determinism. As Clark Glymour puts it, "Determinism requires both the determinateness of quantities and the impossibility of forks in history" (1971, 745). Value determinateness, the third level of determinism, means that physical quantities have exact values. After all, if there are properties whose values are spread out or somehow indistinct, the system would seem insufficiently set, fixed, specified, and determined by the laws of physics. Glymour cites Peirce and Reichenbach as others who regarded value determinateness as a crucial component of determinism, and recently Mark Stone included it in his definition as well: "For deterministic systems, the accuracy of a state description is infinitely refinable, even though any given state description will contain some error" (Stone 1989, 125). Again, this level of determinism can be applied to the universe as a whole to yield the metaphysical belief in the determinateness of the values of all physical properties.

Total Predictability

This last layer of determinism is the idea that the universe is predictable, in principle, by an all-powerful intelligence or computational scheme, given complete information of instantaneous conditions and the complete set of physical laws. This layer is the type of determinism that Popper labels "scientific determinism" and I will call total predictability; namely,

> the doctrine that the state of any closed physical system can be predicted, even from within the system, with any specified degree of precision, by deducing the prediction from theories, in conjunction with initial conditions

---

man points out, the need to control for small influences from outside any small system quickly drives us to the global form of determinism (Earman 1986, 34).

> whose required degree of precision can always be calculated (in accordance with the principle of accountability) if the prediction task is given. (Popper 1956, 36)

Note that this formulation of determinism as total predictability is also used by Prigogine and Stengers to characterize the deterministic worldview of classical physics (1984, 77).[13]

Each of these four layers or levels of determinism is separable; that is, each new level involves additional criteria that are not included in any previous level.[14] Differential dynamics does not imply uniqueness of evolution, for instance. Once we have specified the differential equations governing a physical system it is an open question whether the evolution of the system must follow a unique trajectory. This question takes the mathematical form of inquiring into the existence and uniqueness of solutions to these equations (Earman 1986, 21).

Unique evolution, in turn, does not mean that the system has determinate point values for all variables. As Earman states it (recall that the determinism he refers to here is type B): "Determinism does not presuppose sharpness of values, for we can understand determinism as a doctrine about the evolution of set or interval valued magnitudes as well as about point valued magnitudes" (Earman 1986, 226). This point will prove to be crucial in the discussion of quantum mechan-

13. Additionally, there is Popper's notion of "metaphysical determinism," which says that "all events in this world are fixed, or unalterable, or predetermined. It does not assert that they are known to anybody, or predictable by scientific means. But it asserts that the future is as little changeable as the past" (Popper 1956, 8)—the future is fixed now and for all time, even though the above four layers of determinism may fail to hold. This notion actually seems more concerned with the present truth values of future-tensed statements, but it does illustrate that predetermined is not the same as predestined.

14. This ordering of the four layers is not meant to suggest a rigorous scheme of logical relationships and is partially dictated by rhetorical considerations. While layer A should clearly come first, layers B and C could have been reversed in the "ranking." Also, a different explication of total predictability could lead to its placement elsewhere in the ordering. But the arrangement here appears in Stone 1989 and I adopt it so as to present the four layers in the reverse of the order I follow to argue against them.

ics to follow. Finally, even the combination of the first three layers of determinism does not yield total predictability. Chaos theory presents us with examples of systems that are described by differential equations, follow a unique path of evolution, have determinate values, and yet are not predictable.

Chaos Theory vs. Total Predictability

Chaotic systems scrupulously obey the strictures of differential dynamics, unique evolution, and value determinateness, yet they are utterly unpredictable. Because of the existence of these systems, we are forced to admit that the world is not totally predictable: by any definition of determinism that includes total predictability, determinism is false. Thus begins the process of peeling away the layers of determinism that are not compatible with current physics, impelling us either to revise our definition of determinism or reject it as a doctrine.

Earman, Hunt, and Stone have each convincingly argued that chaotic systems require us either to revise or to discard determinism. Their arguments trade on the fact that total predictability requires that a system not manifest sensitive dependence on initial conditions, for even a system characterized by point values that evolve along a unique trajectory according to straightforward differential equations will be unpredictable if small differences in initial conditions lead to widely separated trajectories later on.

Since no measurement can be made with unlimited accuracy, a chaotic system will be unpredictable even though it satisfies all the other requirements for determinism. Hunt explains that chaotic systems are physically deterministic (that is, they satisfy levels A, B, and C) but not epistemically deterministic (i.e., they are not totally predictable):

> Classical mechanics remains a physically deterministic theory. Every single initial point in phase space is on a path leading to a single later point. But when we admit the impossibility of perfectly accurate measurement,

> epistemic determinism must depend upon the existence of a further relation—continuity—between the paths of [a] single system starting under slightly different initial conditions. It is the assumption of this further relation that is incompatible with the existence of chaotic systems. The failure of epistemic determinism does not therefore undermine the thesis of physical determinism. (Hunt 1987, 132)[15]

John Earman also seeks to defend determinism in the face of the unpredictability of the universe, seeing as the lesson of chaotic systems, "not that determinism fails but rather that determinism and prediction need not work in tandem; for the evolution of the system may be such that some future states are not predictable (at least under Popper's structures) although any future complement than the one fixed from eternity is impossible" (Earman 1986, 9). In other words, sensitive dependence on initial conditions means that we will never be able to tell *which* unique trajectory a system is following, but that does not mean such trajectories do not exist. Classical physics tells us that, given the determinate state of the universe at this moment, all past and future states of the universe are fixed. Our measuring devices and computational skills are not able to find out these other states, because some vagueness is ineliminable in our methods and sensitive dependence will cause any such vagueness to grow, rendering prediction impossible.[16] But that does not mean that the world is not deterministic, if we now are willing to restrict determinism to mean only differential dynamics, unique evolution, and value determinateness.

In other words, determinism (specifically, the layer of

15. Technically, the failure of the continuity condition discussed by Hunt is not the same as sensitive dependence on initial conditions, but both are certainly characteristics of mixing systems, of which chaos theory provides several examples.

16. Mark Stone has rigorously spelled out the algorithmic complexity involved in predicting the evolution of chaotic systems (1989, 127).

unique evolution) says: if two worlds are identical, they will stay identical. This in itself has nothing to do with whether or not we can tell if two worlds (or two systems) really are identical. Notice that Earman is relying on a distinction between the world as it is and the world as we could ever know it. Earman explicitly characterizes determinism as a metaphysical, rather than an epistemological doctrine: "whether it is fulfilled or not depends only on the structure of the world, independently of what we could do or could know of it" (Earman 1986, 7).

Earman thus acts as the defender of determinism, while on the other hand Ilya Prigogine acts as the enemy of determinism in his writing, suggesting the "deterministic worldview" has fallen because of the existence of extremely unstable systems, which are "intrinsically random."[17] His argument may appear somewhat verificationist in that he claims deterministic trajectories are unobservable idealizations for sufficiently unstable systems. Since all our measurements and even our definitions must have some roughness to them, we could never fully specify the state of a system, even in classical mechanics. Since we cannot completely specify the state of a system, and sensitive dependence on initial conditions makes even close approximations rapidly obsolete, Prigogine characterizes deterministic trajectories as illegitimate idealizations.[18]

Here is one way to view this dispute: Earman looks at the form of the equations used in classical mechanics, sees that these models of the world can satisfy the existence/uniqueness condition that he calls determinism, and so concludes that

17. See, for instance, *Order out of Chaos* (Prigogine and Stengers 1984, 177–78).

18. Prigogine's proposal is quite a bit more complicated than this rough sketch might suggest, for it provides a mathematical formalism expressing the "complementarity" between descriptions at the level of individual trajectories and the level of ensembles of systems (1980, chap. 8). Another philosopher of science who works on chaos theory, Robert Batterman, provides a useful formulation of this position, and argues that Prigogine and his colleagues have failed to give a convincing proof of the inadmissibility of exact states (Batterman 1991).

there is no problem. Predictable is not the same as deterministic for the same reason that epistemology is not metaphysics: just because no one could ever use the equations to find that single trajectory the world/system follows does not mean it is not there. And even more than that, the mathematics of the situation tells us that the trajectory is there, for it is built into the formal apparatus of the theory. Determinism is a metaphysical doctrine, a statement about the structure of the universe's coming-to-be. If our theories, our best theories, imply unique evolution, then our theories are deterministic and determinism is not challenged by chaotic systems.

Prigogine might well respond to this by challenging the very distinction between levels B and D of determinism. Such an argument would claim that the distinction between prediction and determinism, like that between methodology and metaphysics, rests on one further split, in this instance the distinction between the theories as they are written down and the theories as they are actually used. Here it is important to pay close attention to the words of Hadamard, who writes that in systems marked by sensitive dependence on initial conditions, "everything takes place, physically speaking, as if the knowledge of . . . [the initial] data would *not* determine the unknown function" (quoted in Earman 1986, 9). How seriously do we take this "as if"? For if everything happens as if the present does not determine the future, determinism is seriously challenged. Joseph Ford lends support to this idea when he notes that even a mathematical proof that a trajectory is unique does not preclude it "from passing every computable test for randomness or being humanly indistinguishable from a realization of a truly random process" (Ford 1987, 3).

If our theories, on paper, speak of a unique trajectory but that trajectory is not and cannot ever be observed or used, and cannot enter into useful practice, then we have run into the question of realism with regard to deterministic trajectories. For in situations marked by sensitive dependence on initial conditions,

> the concept of deterministic evolution along phase space trajectories cannot be defined operationally and hence constitutes a physically unrealizable idealization. Therefore, in dealing with dynamically unstable systems, classical mechanics seems to have reached the limit of the applicability of some of its own concepts. This limitation on the applicability of the classical concept of phase space trajectories is—it seems to us—of a fundamental character. It forces upon us the necessity of a new approach to the theory of dynamical evolution of such systems which involves the use of distribution functions in an essential manner. (Misra, Prigogine, and Courbage 1979, 4–5)[19]

If we actually go wrong in our physics by insisting on using such idealizations, how can we continue to claim that determinism is untouched? Three alternatives are possible here:

1. We can grant Earman that metaphysical determinism, conceived as excluding layer D, remains true even though the deterministic approach to doing science has run out of steam. This move would mean retaining determinism as a cosmological tenet while emptying it of much methodological import and motivational force. Prigogine and Stengers write as though they are willing to make this concession to determinism when they say, "God could, if he wished to, calculate the trajectories in an unstable dynamic world" (Prigogine and Stengers 1984, 272). This alternative would make the "irreducible randomness" of such systems merely an artifact of our epistemic limitations, our inability ever to construct the unique trajectories that must nevertheless be operative.
2. Or we could take more seriously the talk of unique trajectories being an inadmissible idealization. This alternative would involve issuing a challenge to the distinction be-

19. If this use of distribution functions is indeed forced upon us, it would seem that chaos theory can argue against even layer A of determinism.

tween theories on paper and theories in practice and saying, "No, determinism fails because our best theories (as actually used) argue against it." In effect, this means using chaos theory to argue against unique evolution (level B of determinism) as well as total predictability (level D). Such an approach risks becoming naively instrumentalist and too reminiscent of the proposal that we should cease all talk of atoms because we never see them. But this argument has promise, particularly if it could build on the idea that the limitation on predictability imposed by chaotic systems is not simply a practical limitation (see chap. 2). That is, there is no point in insisting that the future is fixed, given the present, if that future cannot be predicted, for if prediction is impossible *in principle,* what more indeterminism could we want? Or, rather, what exactly is the use of a determinism that has been forced to retreat wholly into the realm of the metaphysical, with no methodological cash-value?

3. The approach I favor draws on the implications of quantum mechanics to show that even "on paper," layer B of determinism runs into problems.

Quantum Mechanics vs. Value Determinateness

My argument requires that I introduce the following point, which is a consequence of quantum mechanics: we should either revise our concept of determinism so that it does not include layer C or else reject determinism as false.[20] This layer of determinism, which I call value determinateness, is characterized by the philosopher of science Michael Redhead as the foundation of one possible interpretation of quantum mechanics. This interpretation takes off from the view that "all

20. Clark Glymour writes, "Altogether, I think it almost conclusively established that the quantum theory is not compatible with that aspect of determinism which would require that all physical quantities have precise values at all times" (1971, 749).

observables, in all states, have sharp values" (Redhead 1987, 82). Redhead constructs a careful argument to the conclusion that "by building a few more features" into this view, "we can actually show that this is not a possible way of interpreting, i.e. understanding," quantum mechanics (p. 48).

The argument has two equally powerful components. First he demonstrates that the view of value determinateness, when combined with quantum-mechanical formalism and the additional qualification (known as "locality") that sharp values cannot be changed by changing far-away instrument settings, yields the mathematical relations known as the Bell inequalities.[21] Since predictions derived from these inequalities are contradicted by experiment, either value determinateness, or locality, or quantum mechanics itself must be at fault.

The second component of the argument calls on the Kochen-Specker paradox, which demonstrates that the view of value determinateness, together with certain simple algebraic requirements, yields a logical impossibility. In effect, this means that if you hold to the idea of all observables having sharp values at all times, even an apparently simple algebraic constraint on the way these values "fit together" generates a contradiction. The implication is that we must either reject value determinateness or take an approach that makes you "bound to accept some form of nonlocality" (p. 150). It must be admitted that what these arguments yield is not so much a definitive assault on the idea of determinate values as a convincing portrait of the persistent difficulties such a view brings with it.

Now one may well wonder what determinism can still mean if we live in a world with indeterminate-valued properties, but Earman's point is a good one: determinism requires some amount of determinateness, if only to allow the idea of a change in the state of a system to make sense. But a vision of a universe where systems with somewhat "spread out"

21. On these inequalities, see Bell 1987. Kellert, Stone, and Fine 1990 has a short derivation of these relations (pp. 97–99).

properties evolve along uniquely determined paths retains coherence.[22] The question remains, however, what would happen if chaotic dynamics were at work in such a universe.

Chaos Theory and Quantum Mechanics vs. Unique Evolution

Imagine a simple system in a phase-space of two dimensions—say a particle moving in a one-dimensional potential well. In the classical picture, the state of the particle at one instant t_0 is given by a point (x_0, p_0) in state space (see fig. 7). Determinism (incorporating layers A, B, and C) says that given this point, plus the relevant boundary conditions and all physical laws, the particle's trajectory is fixed for all other times. That is, if there is another system identical to this one, and if it is also in state (x_0, p_0) at time t_0, then both systems will be in the same state at all other times $t_0 \pm \Delta t$. One point, at one instant, suffices to establish all the past and future behavior of the system.

Now, as my physics professor Robert Adair would say, let's turn on quantum mechanics. Our particle can no longer be assigned a straightforward pointlike state (x_0, p_0). Instead, the state of the system is given by specifying a function Φ_0 at t_0—a specification that gives us all the information there is about its physical state. The uncertainty relations tell us that specifying this state-function suffices only to associate the particle with a "patch," not a point.[23] This patch will have an area of not less than $\Delta x \Delta p = \hbar/2$ (see fig. 8). Although this patch may be very narrow (very small Δx) or very short (very small Δp), its total area will be at least $\hbar/2$—the idealization of a mathematical point is not just practically difficult to reach or operationally indefinable but theoretically precluded.

Earman argues convincingly that this does not ipso facto defeat determinism. That is, the universe may well evolve

22. Fine 1971 and Teller 1979 work out such a vision in terms of set- and interval-valued properties.

23. Thanks are due here to Linda Wessels for helping me to speak less misleadingly about the quantum-mechanical situation.

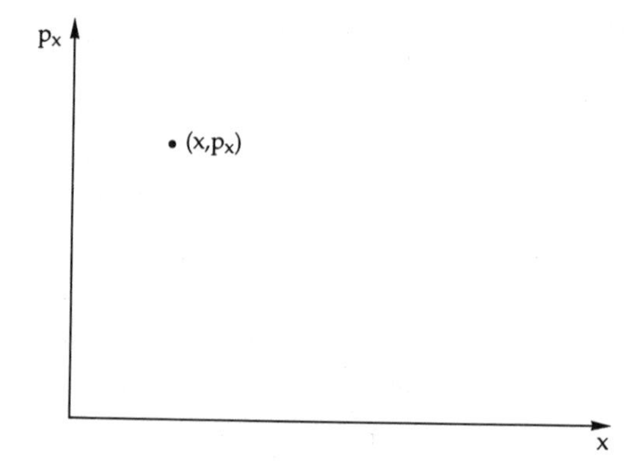

Fig. 7. A classical system

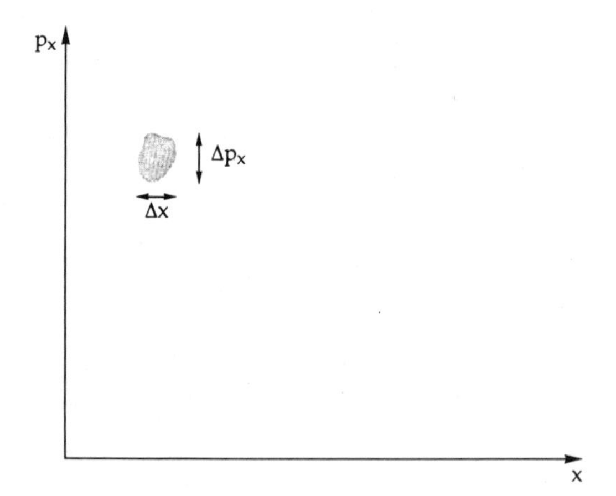

Fig. 8. A quantum system

along not a one-point-thick trajectory but a slightly blurry trail with some nonzero "thickness." Alternatively, two systems each characterized by Φ_0 and an associated "patch" at t_0 will evolve to be described by the same Φ_t at $t > t_0$; the patches associated with the two systems remain identical at all times. That is, the objects in our universe never have perfectly sharp positions, but their history of spread-out positions is nonetheless fixed for all time by fixing them at one instant. Here is the heart of this particular rescue of determinism from quantum mechanics suggested by Earman: by isolating the layer of unique evolution and making it the essence of determinism we can continue to maintain that a quantum-mechanical world is deterministic.

What happens to determinism when we now turn on chaos theory in conjunction with quantum mechanics? Recall that systems characterized by sensitive dependence on initial conditions are such that, given a point in phase-space, there is always another nearby point that will follow a widely divergent trajectory. The criterion of chaos is that two systems that are both located within the same small patch of state space can diverge greatly in a relatively small amount of time: "the common feature of dynamical systems having a suitably high degree of instability is that each finite region of state space, no matter how small, contains points that move along rapidly diverging or qualitatively distinct types of trajectories" (Misra, Prigogine, and Courbage 1979, 4).

Consider the following argument:

> Quantum mechanics says a one-particle system cannot be said to have a pointlike state in state space: the totality of physical information about it suffices only to identify it as a patch of finite area with a lower bound on its size.
>
> Chaos theory says that two otherwise identical chaotic systems with slightly different initial conditions will eventually diverge greatly, no matter how small the initial difference.
>
> Therefore:
>
> Two physically identical chaotic systems with identical boundary conditions and laws and with their one particle in the same physical state at t_0 can be in different states at $t > t_0$. That is, determinism as uniqueness of evolution fails to hold.

Here it must be admitted that I am equivocating about what are to count as "initial conditions" and even "the same state": do two systems specified by the same state function and localized within the same patch in state space have the same initial conditions or not?[24] Indeed, the whole project of tacking chaos theory onto quantum theory may seem illegitimate, since they differ so widely in the mathematical expressions they use. But the scientific study of quantum chaos is a very new field, so new that there is still some question as to whether it even exists. I am at a loss to adjudicate competing claims about the possibility of chaotic behavior in quantum-mechanical systems. Earman states that evolution according to the Schrödinger equation cannot exhibit sensitive dependence on initial conditions or any stronger version of instability (Earman 1986, 201). But on the other hand, Prigogine and George write that examples of unstable and intrinsically random quantum systems are well documented (1983, 4594).

In fact, the most fruitful way to regard "quantum chaos" may be to follow the formulation of one of the leading researchers in this emerging field, Martin Gutzwiller, who sees it as an inherently transitional field of investigation (1992). By this account, quantum chaology seeks to understand the relation between chaotic behavior in classical systems and nonchaotic behavior in quantum systems, but not by looking for actual cases of chaos in quantum systems. Instead, "quantum chaology" looks for the emergence of chaos in what are known as "semiclassical" systems, which incorporate some limited quantizing principles. Such a study also focuses on quantum systems at energies high enough to begin to exhibit behavior more typical of macroscale systems

24. One possibility, suggested by John Earman, is to look at the evolution of $\langle x \rangle$ and $\langle p \rangle$, the expectation values of the particle's position and momentum, instead of the evolution of the state function itself. Ehrenfest's theorem tells us that these expectation values must satisfy the equations of classical dynamics; see, for example, Cohen-Tannoudji, Din, and Laloë 1977, 242–43.

(see also Berry 1987). The field of quantum chaos is in such a state of growth and flux, however, that it would perhaps be premature to attempt much more of a characterization than this.

In light of these considerations, I will reformulate the argument of this section so that it does not crucially depend on the outcome of investigations in quantum chaos. This new formulation sees quantum mechanics as saying, "systems can never be associated with a patch smaller than $\hbar/2$" and chaos theory as saying, "a system with sensitive dependence on initial conditions that can be located only within a patch of finite size will, after some time, be able to be located only within a patch of much larger size." So two identical systems, each localized as a very tiny patch of allowable state space, will eventually both be localized only as a huge patch—perhaps even the entire allowable region. Can we still maintain that the world is deterministic?

I contend that we cannot. Chaotic dynamics will take the tiny indeterminacies of quantum-mechanical systems and stretch them into huge variations, dilating the smallest patch until, at some sufficiently distant time in the future, almost anything is possible.[25] Given the current state of the universe, the future will be fixed, but fixed only within a huge spectrum of distinguishable possibilities. As John Earman admits, "Determinism does seem to presuppose some minimal amount of determinateness; if the world were entirely a froth of potentialities with no magnitudes having determinate values, point or interval, one would be at a loss to say whether determinism held or failed" (1986, 226). Chaos and quantum theory lead us to just such a vision of the universe as a congeries of interrelated but open possibilities, foaming forth in its infini-

25. In some sense, a chaotic system can thus act as a measurement apparatus for a quantum system, coupling with that system and amplifying unobservable micro-changes to the level of everyday effects. Recall that the device to kill Schrödinger's cat had to be poised with enough instability to be sensitive to very small changes.

tude. Determinism is not so much proven false as rendered meaningless.[26]

One final way to formulate this argument proceeds by casting level B of determinism as an if/then statement. The tenet of unique evolution means that *if* there were two identical worlds at time t_0 *then* they would be identical at all other times. The challenge to this form of determinism can now be cast as a dilemma:

I. Does "two identical worlds" mean "all particles have the same position, momentum, etc., even to an infinite number of decimal places"? If so, then quantum mechanics guarantees that the antecedent condition can never be met; such a determinate-valued universe cannot be legitimately formulated, so determinism tells us nothing about our world.
II. Or does "identical" mean "the same, so far as physics can specify"? If it does, then two systems that start with the same small patch of area $\hbar/2$ in state space are the same in all physically relevant respects. But if they are chaotic systems they can and will evolve in such a way that the patch of vagueness will grow to a size where judging them the same or different loses its meaning.[27]

In other words, chaos theory together with quantum mechanics invites us to participate in the following imaginative exercise: picture another universe, created at just this very moment, which is physically identical to this one. Let the physical descriptions of the two universes be identical (never mind that such descriptions could never be written down, the point is that the two worlds agree on "all physically relevant

26. A similar argument has been made by the philosopher Jesse Hobbs (1991), formulated in terms of micro-level indeterminateness "scaling up" to the macro-level. Hobbs seems to imply that the link between chaotic behavior and indeterminism rests on the self-similar structure often manifested by chaotic behavior, however (p. 157).

27. Thanks to Jennifer Church and Doug Winblad for helping me to become clear on this point.

properties"). Now imagine watching them unfold in time (again, never mind where you are supposed to be watching them from). After a while, they will be noticeably different—an atom will decay in one and not in the other, a hurricane will strike Florida in one and not in the other, two asteroids will collide in one and not in the other. Determinism fails. Now attend to your feelings as you contemplate the fact that the historical evolution of the physical universe is fundamentally *open*. There may be a queasy feeling of panic and delight—delight in the unfixed possibilities, panic that (to paraphrase the T-shirt) stuff happens. It just happens.

Differential Dynamics as Local Determinism

Recall the motivational concerns Earman describes, the feeling that if determinism is false then physics will have to appeal to divine intervention or blind chance. Must the defeat of the layer of unique evolution force us into this Pynchonesque dilemma? I contend that the methodological injunction generated from the remaining layer A, differential dynamics, is enough to prevent this. The exhortation to use nonstochastic differential equations to understand physical change urges us to seek local connections between events without concern for the global property of unique evolution. This layer of determinism, which I shall call local determinism, is all that we need to encourage the continued progress of mathematical physical science.

The project of local determinism is to construct differential dynamical systems. Recall that such systems must describe the evolution of states, not probabilistic distributions or ensembles of states. Furthermore, they must describe this evolution with differential equations that make no explicit reference to chance—the equations must not include possible branchings or inherently stochastic elements. There is no doubt that the research into chaotic systems takes place wholly within a context of local determinism. As I will describe in the next chapter, chaos theorists seek to understand complex behavior by building models that obey precisely the strictures of differential dynamics.

But even this local determinism is not without its problems. First is the challenge to the project of studying the evolution of single states. In an indeterminate world, where any margin of indeterminacy can make a great deal of difference down the road, single states may in fact be an inadmissible idealization. As Holloway and West write, "the point of view of a segment of the scientific community has shifted more towards the perspective that purely deterministic behavior is an illusion and only distributions, albeit very narrow in some cases, have physical significance" (Holloway and West 1984, v). The second problem is whether stochastic evolution is intrinsic in quantum-mechanical systems. In the guise of questions about what von Neumann called "Type II evolution," this is precisely the quantum measurement problem.[28]

Chaos theory, together with quantum mechanics, leaves us with a much-narrowed conception of determinism at work in the world. Considering some of the motivations for determinism sketched earlier, this result may seem to suggest that we are adrift in a sea of indeterminate flux, lost without total predictability and ultimate explanations. Indeed, the motivations for interest in the metaphysical tenet of determinism share deep connections with epistemological concerns about what form knowledge should take: totalizing predictions phrased in an atemporal language of certainty, not contingency. But I will conclude with the advice of Wesley Salmon, who argues that even in the face of indeterminism we need not despair of understanding the physical world: "If indeterminism is true, it does not follow that there are events that are incapable of being explained" (Salmon 1971, 343). Many recent discussions of scientific understanding focus on the clarification of explanations in the face of probabilistic phenomena. In my next chapter I will consider just how research into chaos provides us with an understanding of indeterministic events.

28. See, for example, Jammer 1974, 474–82.

4

On the Way to Dynamic Understanding

> It had been some time since Gregorovius had given up the illusion of understanding things, but at any rate, he still wanted misunderstandings to have some sort of order, some reason about them.
>
> —Julio Cortázar (1966)

The Uses of Chaos

Contemporary science has become the place to turn for the legitimation of unconventional and, occasionally, outlandish claims. The past few years have seen chaos theory used to the hilt for both worthwhile interdisciplinary cross-fertilization and fashionable rhetorical co-optation. In Chicago, Jean Baudrillard proposes a fractal model of the postmodern self, while in New York a talk is advertised called "Tantra, Sufism, and Chaos Theory." Chaos theory is portrayed as paralleling and confirming the insights of literary theory (Hayles 1990) as well as Taoism (Briggs and Peat 1989).

In addition to the cognitive authority of mainstream science, chaos theory has much to offer. It has a neat name. It generates pretty pictures. It represents the very latest thing in science. It is relatively accessible. But besides these trappings, it must be admitted that chaos theory challenges many of our presuppositions and makes us think differently about the world. All of these reasons are important, but it is this last one that is crucial for fueling the "gee-whiz" aura surrounding nonlinear dynamical systems theory.

The reigning champion for inspiring wildly speculative associations with the sciences is quantum mechanics. A rash of books have purported to show connections between quantum theory and any number of Eastern religions, for instance.[1] There is a relatively simple explanation for this phenomenon: hardly anyone in the West understands either. Faced with an incomprehensible and counter-intuitive physics and an incomprehensible and counter-intuitive religious tradition, the Western mind somehow figures that they both must be saying the same thing.

But surely it is ethnocentric to think that all challenges to the Enlightenment worldview are the same, or parallel, or even potential allies. Is it not somehow arrogant to assume that whenever something assigned to the Outside of our culture (the physical world, the East, etc.) challenges our presuppositions, the challenges must have interesting connections between them? Granted, multiple crises now confront the world-picture of Modernity. And chaos theory also presents problems for traditional views of the natural world and of scientific understanding. But before assimilating all these challenges into one grand message for Western Culture, the first task must surely be to gain a clear sense of just what these challenges are.

One of the most intriguing suggestions made by recent considerations of chaos theory is the idea that chaos theory invites us to revise our notions of scientific understanding; science must now be seen as holistic, decentered, or dialogic. My goal in this chapter will be to answer the question, Just what kind of understanding does chaos theory give us? as a way to begin thinking about possible revisions of our philosophical accounts of scientific understanding. I will proceed by considering possible answers to three consecutive questions: What does chaos theory give us understanding of? In what way does this understanding arise? And, What sort of

1. See, for instance, Fritjof Capra's *Tao of Physics,* Gary Zukav's *Dancing Wu Li Masters,* and Ben Toben's *Space-Time and Beyond.*

understanding is it? Alternatively, what follows is a preliminary investigation into the object, the method, and the character of the understanding provided by chaos theory.

In answering each of these questions, I will consider various philosophical accounts of understanding, paying attention both to how they can help illuminate the case of chaos theory and to the difficulties this case raises for them. These considerations lead to the following characterization: chaos theory provides an understanding of the appearance of unpredictable behavior by constructing models which reveal order. The "interesting," speculative connections between this understanding and other, marginalized traditions I will leave for others, with one exception. For I mean to show that it is helpful to think of chaos theory as revealing order in natural processes, where "order" is understood along the lines suggested by feminist historian and philosopher of science Evelyn Fox Keller (1985). The connection with feminist philosophy of science will then be more explicitly thematized in chapter 5.

Understanding and Explanation

Before even considering what form of understanding chaos theory provides, we must acknowledge the existence of an open question as to whether it provides any understanding at all. Perhaps, as some have suggested, chaos theory is merely a "new toy" for scientists to play with. Despite the confident tone of chapter 1, there is no universal agreement on exactly what "chaos theory" is; this ambiguity creates the problem of establishing that it does in fact provide scientific understanding. After all, if chaos theory is best conceived as a "new microscope," it would seem to give us new observations or even new phenomena, but no understanding at all.

One strategy for dealing with this problem is to rely on the testimony of scientists. In the introduction to one recent volume we read, "many features of chaos in low-dimensional systems are widely understood both from the theoretical side and from the experimental point of view" (Livi et al. 1988, vi). Roderick Jensen writes that we have "a clear understanding" of chaos in classical systems (Jensen 1987, 169; see also Kadanoff 1985, 39). The burgeoning literature of nonlinear dynamical systems theory is full of claims that recent advances have increased our understanding, even if behaviors such as full-scale turbulence are not yet fully understood.

Some may not be satisfied with this strategy of letting scientists enter the debates over when something gives us understanding and when it does not. There are indeed some good reasons to be uncomfortable with disallowing *any* critical examination by nonscientists of what should count as scientific understanding. But allowing scientists to enter into these disputes should not be read as immediately excluding philosophers from them. In light of these concerns, I make the following suggestion: let us initially construe "chaos theory gives us understanding" to mean nothing much more than that it helps us to think about and respond to some aspects of the world in interesting or useful ways. That is, begin with an extremely generous and inclusive sense of "understanding," and proceed to clarify just what type of understanding is operating here. In this way, we can let the examination come before the judgment.

One final comment is in order before looking at chaos theory in comparison with philosophical accounts of understanding. It must be noted that these accounts were promulgated first and foremost as philosophical explications of the notion of scientific *explanation* and not understanding per se. Some may wish to suggest that it is not fruitful to investigate the understanding provided by chaos theory by using accounts which aimed instead at explanation. In this matter, I will follow two practitioners of the explication of explanation, Wesley Salmon and Philip Kitcher, who believe that a philosophical account of explanation provides insight into the nature of scientific understanding (Salmon 1989, 135; Kitcher 1989, 419). Philosophical accounts of scientific understanding have indeed focused on the concept of explanation in recent years, and so it is important to see how such accounts can help make sense of chaos theory. It is also important to evaluate the adequacy of these accounts in the light of recent work in the sciences.

The Object of Understanding

One of the keenest difficulties in comparing philosophical accounts of explanation with the understanding provided by chaos theory results from a mismatch between the item to be explained in the former and the item to be understood in the latter. While philosophers commonly address scientific explanations of such things as "phenomena," "facts," or "events," (cf. Salmon 1989, 4, 5, and 8), chaos theory usually studies such things as behaviors, patterns, or bifurcations.[2]

Much of this discrepancy arises from the different types of questions involved: standard philosophical accounts characterize scientific understanding as arising from an accumulation of explanations which answer "Why questions" (sometimes expanded to include "How possibly questions"). But chaos theory often answers a different kind of question: a "How question." In studying various physical behaviors through the use of simple mathematical systems, the central puzzling questions include How does extremely complicated behavior come to occur in nature? How does it happen that some physical behavior is completely unpredictable? And How do orderly patterns persist amid apparent randomness? These question areas comprise the object of understanding for chaos theory, which I will delineate further in this section. Stated briefly, chaos theory enables us to understand how unpredictable behavior appears in simple systems.

2. A further observation, by Mark Stone, is that while philosophical accounts typically cast the item to be explained as a linguistic entity, researchers in chaos theory more often will point to a computer graphics display and say, "this is what we need to understand."

Describing the object of understanding in this way involves a deliberate equivocation: there are two ways to read the phrase "how unpredictable behavior appears," and these two readings correspond to two aspects of our understanding. In the first place, chaos theory allows us to understand "how unpredictable behavior appears" in the sense of "how does it come to be that simple systems display such complicated behavior?" Here, we are given an account of the way limits to predictability arise and unpredictability emerges. In the second place, chaos theory lets us understand "how unpredictable behavior appears" in the sense of "what does this behavior look like?" Here, we are given an account of how intelligible patterns persist after the onset of unpredictability. In *Order within Chaos* we read the following succinct presentation of the object of understanding for the theory of chaotic behavior in dissipative systems: "The ultimate goal is to understand the origin and characteristics of all kinds of time-evolution encountered, including those which at first seem totally disorganized" (Bergé, Pomeau, and Vidal 1984, 102).

The Transition to Chaos

The mathematical study of the transition to chaos addresses the first of these aspects of the object of understanding. As Francis Moon writes, "A great deal of the excitement in nonlinear dynamics today is centered around the hope that this transition from ordered to disordered flow may be explained or modeled with relatively simple mathematical equations" (Moon 1987, 3). For large-scale, unbounded fluid flow, an understanding of the transition to chaos (or turbulence, in this case), is still being hoped for. For many related, but restricted, systems, the situation is much brighter. Once the "route to chaos" is determined, a researcher can often foresee the value of the control parameter at which the system will manifest unpredictable behavior. Furthermore, the Lyapunov exponent gives a quantitative measure for the rate of decay of predictability. As mentioned in chapter 2, given the initial conditions of a system to some specified accuracy and the Lyapunov exponent, one can calculate a reliable estimate of how long our quantitative predictions about the system will remain worthwhile. This ability to ascertain in advance the onset and intensity of chaotic behavior led one meteorologist to comment that weather forecasters are now reduced to "predicting the limits of their predictions."[3]

Here we again encounter the disheartening effect the discovery of chaos can have: it seems as if all this new science can do is feel out the walls of its prison cell. But chaos theory allows us to account for the limits to predictability we encounter; this means we can understand how these limits arise, which enables us "to analyse the way in which chaos settles in via Lyapunov exponents or the way in which unpredictability appears" (Bergé, Pomeau, and Vidal 1984, 267). The widespread appearance of systems with sensitive dependence on initial conditions means that this is "a world where small causes can have large effects, but this world is not arbitrary. On the contrary, the reasons for the amplification of a small event are a legitimate matter for rational inquiry" (Prigogine and Stengers 1984, 206). Chaos theory provides a way to understand how unpredictability happens; it tells us the way limits to predictability arise in simple systems.

Characterizing Chaos

The ability to understand how unpredictable behavior appears also includes an account of the ways we can in fact make predictions about chaotic systems. One of the apparent paradoxes of chaos theory is that a scientific study of unpredictable systems actually has significant predictive power. This paradox will be resolved later by clarifying the difference between quantitative and qualitative predictions, but for now

3. Tim Palmer, quoted in *New Scientist,* November 19, 1988, p. 56.

it is important to note that chaos theory lets us understand how predictable large-scale or long-term patterns appear in behavior which is nonetheless unpredictable in detail. In this respect, chaos theory bears a certain resemblance to statistical sciences of physical, biological, and social systems. The invention of techniques for statistical analysis revealed orderly patterns (such as the "normal" distribution) in the apparently random behavior of heated gases, animal populations, and undeliverable letters (Gigerenzer et al. 1989). But while statistical techniques analyze averages over a large ensemble of systems, the techniques of nonlinear dynamics work on single systems or families of related systems. The "order" found in a system with chaotic dynamics has little in common with the "orderly" distribution of molecular velocities in a gas, for instance.

Research into chaotic systems often concentrates on characterizing the behavior with the use of such quantitative measurements as the fractal dimension of the attractor or the Lyapunov exponent.[4] The method of reconstructing an attractor for a system from a series of measurements of one variable is often the first step in studying the geometric properties of the behavior of a system (see chap. 1). Such investigations seek to develop a "taxonomy" of chaotic behavior, a way of classifying and cataloguing these systems which at first glance may all seem to be nothing more than expensive white-noise generators.

Once there is a way to describe and differentiate varieties of chaotic behavior it can make sense to investigate how this behavior changes when parameters of the system are changed. Characterizing the patterns of behavior thus makes it possible to study the way a system responds to external influences: if there were no way to distinguish between different types of complicated unpredictable behavior, it would be meaningless

4. For examples of scientists who describe the work of chaos theory in terms of characterizing dynamical behavior, see Bergé, Pomeau, and Vidal 1984, 146, and Jensen 1987, 179.

to ask about how the system changes. Here again, such research is not devoted merely to being able to foresee the system's response to an alteration, but to give an account of the way it changes. This understanding, it has been suggested, may eventually lead to our being able to cure certain "dynamical diseases" that occur in organisms whose internal regulatory systems wander into the wrong range of parameters (Albano et al. 1986, 235 and 238).

It may seem that little of importance rests on whether chaos theory allows one to understand *how* unpredictable behavior appears or *why* it appears. But emphasizing the attention these researchers give to "the way things happen" serves to underscore that they address themselves to an investigation of overall patterns of behavior as opposed to individual events. Certainly chaos theory is sometimes used to answer "why questions," as in explaining "why are there gaps in the asteroid belt at certain definite distances from Jupiter's orbit?" (See Wisdom 1988, 418.) But even these explanations are framed in terms of describing the way these phenomena came to be.

The Method of Understanding

The next question concerns how chaos theory goes about providing an account of the appearance of unpredictable behavior. At issue here is the method of understanding: In what way does chaos theory give us understanding? By what method, by what means? And the answer is: by constructing, elaborating, and applying simple dynamical *models*. The activity of building and using these models has three important characteristics which I shall deal with in turn: the behavior of the system is not studied by reducing it to its parts; the results are not presented in the form of deductive proofs; and the systems are not treated as if instantaneous descriptions are complete. In what follows I refer to these three methodological aspects of chaos theory as holism, experimentalism, and diachrony.

Models

Researchers in chaos theory consistently describe their work in terms of modeling methods. For instance, one scientist characterizes the way to study the onset of hydrodynamic turbulence as follows:

> To understand a complicated phase transition—i.e., a change in behavior of a many-particle system—choose a very simple system which shows a qualitatively similar change. Study this simple system in detail. Abstract the features of the simple system which are "universal"—that is, appear to be independent of the details of the system's makeup. Apply these universal features to the more complex problem (Kadanoff 1985, 29).

Here we see a powerful example of empirical evidence from the sciences working to support a particular position in the philosophy of science. In this case, that position is known as the semantic view of theories, expressed in Ronald Giere's injunction that "When approaching a theory, look first for the models and then for the hypotheses employing the models. Don't look for general principles, axioms, or the like" (Giere 1988, 89).[5] But the argument of this section will not be only that a focus on models better describes what chaos theory is. Rather, looking at models better describes the method chaos theory employs to provide understanding. Better, that is, than philosophical accounts which portray science as proceeding by methods which are microreductionist, deductivist, and synchronic.

The heart of the semantic view is the notion of a model: an idealized system which is defined by a set of equations. A useful example of such a model is the simple harmonic oscillator, an abstract entity which is defined by the statement that it satisfies the force law $F = -kx$ (Giere 1988, 78). While no

5. The semantic view has been propounded by several philosophers of science, including Frederick Suppe (1977) and Bas van Fraasen (1980).

actually existing physical system obeys this law exactly, the simple harmonic oscillator and its numerous variants comprise a family of models which has been applied to the study of widely disparate types of behavior.

The application of these models to actual physical situations proceeds by the elaboration of theoretical hypotheses: statements which assert a relationship of similarity between a model and a particular system or class of systems (Giere 1988, 80). He provides the example of the Newtonian model of a system with two point-particles experiencing gravitational force: the theoretical hypothesis in this case asserts that this model is similar to the earth-moon system with respect to their positions and momenta. The mathematical model, given the relevant initial conditions and values for the masses and the constant of gravitational attraction, will agree in its values for positions and momenta to within some specifiably small margin of error.

A theory, on this account, comprises two elements: "(1) a population of models, and (2) various hypotheses linking those models with systems in the real world" (Giere 1988, 85). Presentations of chaos theory, in textbooks and seminal articles, consistently follow the scheme of displaying an array of useful models and ways to apply them to actual situations. Works such as *Nonlinear Dynamics and Chaos* (Thompson and Stewart 1986), *Order within Chaos* (Bergé, Pomeau, and Vidal 1984), and *Chaotic Vibrations* (Moon 1987) offer parallel treatments of what have now become standard models: the logistic map, the Lorenz system, and the Henon attractor, among others. Hardly any full discussion of chaos theory will neglect these models, which serve as "paradigms" according to Francis Moon, or, better, as "exemplars."

In the next section, I will discuss the specific theoretical hypotheses which, together with these models, make up chaos theory. Spelling out exactly what kind of similarity is asserted to hold between these models and actual physical systems will make clear the character of the understanding these models give us. But in the remainder of this section I will deal with the method chaos theory uses for constructing its models—a method which is neither microreductionist nor deductive nor ahistorical.

Holism

A microreductionist method seeks to gain understanding of a system by the time-honored method of analysis: breaking the system into its constituent parts and searching for the lawlike rules that govern their interaction. Robert Causey uses the term microreduction to signify "an explanation of the behavior of a structured whole in terms of the laws governing the parts of this whole" (Causey 1969, 230). The prodigious successes which modern technology has achieved in manipulating our physical environment bear witness to the power of this method, and many sciences seek to emulate the approach of subatomic physics which provides a rich understanding of matter on the tiniest of scales by breaking it into successively smaller components.

In classical physics, the usefulness of the microreductionist method was borne out by the proliferation of models with the property of integrability. An integrable model promises a mathematical expression in terms of elements which experience no interaction (Prigogine and Stengers 1984, 72). With a sufficiently clever change of variables, an integrable system of interacting parts can be mathematically transformed into a system of mutually isolated parts under some constraints. Exact solutions for such models can often be found, which yield a formula with comprehensive predictive power. The promise of such predictive power helped to motivate a commitment to the microreductionist method.

But consider a simple system like a double pendulum: one simple pendulum swinging from the end of another pendulum which in turn is driven at a constant frequency by a force such as an escapement mechanism. (Such a device has a precise analog in electrical circuitry.) We can write down the equation which governs this system, but since it is nonlinear and nonintegrable, we cannot reduce the system to two separate oscillators. For certain values of the driving frequency, the device will oscillate in a chaotic, unpredictable fashion (Moon 1987, 284–85). But knowing the parts making up the system and the equation governing their interaction does not tell us what we want to know: it does not help us understand how the chaotic behavior sets in, or what kind of attractor characterizes it, or how the system will respond to changes in the driving frequency.

Understanding this behavior requires the use of the new mathematical techniques of reconstructing attractors and looking at surfaces-of-section and building first-return maps and computing fractal dimensions. None of this information drops out of our knowledge of the governing equations the way knowledge of eclipses followed directly from Newton's model of the solar system.

> The hope that physics could be complete with an increasingly detailed understanding of fundamental physical forces and constituents is unfounded. The interaction of components on one scale can lead to complex global behavior on a larger scale that in general cannot be deduced from knowledge of the individual components. (Crutchfield et al. 1986, 56)

When dealing with chaotic systems, Mitchell Feigenbaum says, "you know the right equations but they're just not helpful. You add up all the microscopic pieces and you find that you cannot extend them to the long term. They're not what's important in the problem. It completely changes what it means to *know* something" (quoted in Gleick 1987, 174–75).

This last sentence is quite provocative and will be dealt with later. For now, it is important to clarify that chaos theory argues against the universal applicability of the method of microreductionism, but not against the validity of the philosophical doctrine of reductionism. That doctrine states that all properties of a system are reducible to the properties of its parts, where the reduction may be spelled out in terms of logical equivalence, supervenience, or the like. Chaos theory gives no examples of "holistic" properties which could serve as counterexamples to such a claim. No researcher is likely to say, for instance, that the positive Lyapunov exponent of a system is a property of the system as a whole which is ontologically distinct from the properties of all its parts.

In contrast, microreductionism as a methodological injunction asserts that it is always appropriate to seek to understand the behavior of a system by trying to determine the equations governing the interactions of its parts. The fruitfulness of chaos theory militates against this creed; to quote Feigenbaum again, to engage in "the business of writing down differential equations is not to have done the work on the problem" (in Gleick 1987, 187).[6] To say this flies in the face of much of physics as it is currently taught, where once the instructor has written down the equations describing the forces at work he or she may typically turn to the class and say "now we are done with the physics of the situation." In contrast, consider James Yorke's assertion that in studying chaotic systems, "Sometimes you can write down the equations of motion and sometimes you can't. Our approach is to ignore the equations and carry out the analysis without knowing them" (Yorke 1990).

Experimentalism

The belief that the microreductionist method is universally applicable in studying physical behavior traditionally forms part of a two-stage portrait of the method of gaining under-

6. Please note that this distinction between reductionism as a methodology and as ontology may be ultimately unsupportable. For if there are aspects of a system that simply cannot be understood, even in principle, in terms of its constituent parts, what is the force of insisting that these "global" aspects are nonetheless ontologically dependent on the system's constituents? As in the case of determinism, a metaphysical tenet utterly lacking in methodological import is empty.

standing. The first stage is reductive analysis and the second stage is the construction of a deductive scheme which yields a rigorous proof of the necessity (or expectability) of the situation at hand. And just as chaos theory makes little use of microreduction, it has little place for strictly deductive inferential schemes.

The conception of chaos theory as a family of models already makes room for the insight that there need be no deductive structure which generates all these models from some simple set of propositions. The semantic view allows for the possibility of inferential links between the models, but does not require them. Indeed, to the extent that one model is developed from another by judicious approximation, the connection between them is "*not* a matter of purely mathematical, or logical *deduction*" (Giere 1988, 71).

My point here is not to argue that chaos theory pursues understanding by means other than the construction of deductive structure; it is all too evident that many sciences proceed quite well without any rigorous deductive apparatus. But some philosophical accounts maintain that science pursues understanding by helping to indicate how to fill in deductive schemes which may never be actualized or even mentioned by the scientists themselves. I have in mind Peter Railton's notion of the "Ideal Explanatory Text" and Philip Kitcher's theory of explanation as unifying deductive schemes.[7]

In contrast to these accounts, I would contend that nonlinear dynamics does not provide understanding by helping us fill in overarching inferential patterns. Chaos theory often bypasses deductive structure by making irreducible appeals to the results of computer simulations. The force of "irreducible" here is that even in principle it would be impossible to deduce rigorously the character of the chaotic behavior of a system from the simple equations which govern it.

The difficulty of deriving rigorous results about the simple models studied by chaos theory is notorious. Even such an

7. See Railton 1981 and Kitcher 1989.

exemplar of chaos as the Lorenz system has never been strictly proven to exhibit sensitive dependence on initial conditions.[8] In the face of this difficulty, researchers regularly turn to what they call "numerical experiments," that is, the use of a computer to simulate the behavior of an abstract dynamical system by numerically integrating the equations of motion. In his crucial paper announcing a criterion for the expected onset of chaos in Hamiltonian systems, Chirikov states that he is following the great mathematician Kolmogorov in holding that, "it is not so much important to be rigorous as to be right. A way to be convinced (and to convince the others!) of the rightness of a solution without a rigorous theory is a tried method of science—the experiment. . . . In the present paper we widely use the results of various numerical 'experiments'" (Chirikov 1979, 265).

If this recourse to numerical experiments represented only a matter of convenience in the face of bothersome mathematical difficulties, chaos theory would present no challenge to deductivist accounts of the method of gaining scientific understanding. But here we encounter no mere practical difficulty but an impossibility of the type discussed in chapter 2. To study the behavior of a dynamical system we must look at the orbits of the system, and numerical integration allows us to examine the orbit which passes through some given point in the state space of the system. But in a chaotic system, which manifests sensitive dependence on initial conditions, our simulated orbit is guaranteed to diverge wildly from the actual orbit passing through this point unless our means of mathematical representation and computation are infinitely accurate.

Because we are finite beings and all our computational re-

8. Computer simulations of the Lorenz system allow one to compute the Lyapunov exponent, for instance, but this does not constitute a proof of sensitive dependence (Hirsch 1985, 191). So far, there is no rigorous deduction of the existence of a chaotic attractor for this system of equations (Robinson 1989, 495).

sources are definitionally characterized by finitude, we can never construct an actual orbit of a chaotic system. But an important theorem, called the shadowing theorem, provides that we can nonetheless construct an orbit that is always as close as we need to some actual orbit. Still, "the only humanly feasible way to obtain this locally accurate orbit is with a computer or with some other device of the same name. Hence the computer is no longer disjoint from analysis" (Ford 1986, 47).

In studying simple mathematical models, chaos theory sometimes produces results which cannot even in principle be logically derived from the equations which define these models. So the study of chaos provides understanding through the use of a modelling method which is neither microreductionist nor deductivist.

Diachrony

The general methodology of physics is marked by synchrony: the pursuit of understanding in terms of the properties of instantaneous states. This ahistorical tendency is represented in dynamical systems theory as the characterization of the state of a system solely in terms of the way the system is at one moment. Physics considers that we know everything relevant about a system if we know everything about it at one point in time.

But consider a simple system consisting of a mass on a spring with a nonlinear force function (a "hard" spring) being driven by an external periodic force with a frequency w. As we increase the driving frequency to w_a starting from zero, the spring will start to oscillate with a small amplitude. This behavior is represented as point (a) on the diagram below.[9] Increase the driving frequency to w_b and the system will swing with greater amplitude, the familiar phenomenon of reso-

9. This discussion is drawn from Abraham and Shaw 1982, 144–46.

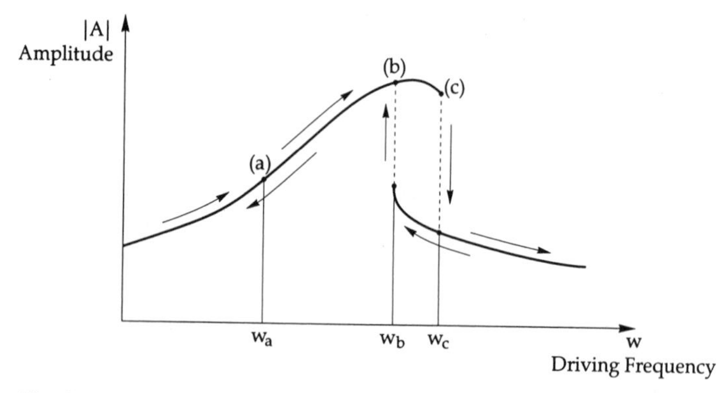

Fig. 9. Hysteresis. From Abraham and Shaw 1982

nance (point (b)). But if the driving frequency continues to be increased, eventually we reach a value w_c after which a drastic decrease in the resultant amplitude occurs (point (c)).

The strange phenomenon known as hysteresis manifests itself if we now decrease the driving frequency from w_c to w_b. The system will suddenly start oscillating with a drastically increased amplitude, but this leap takes place at a different value of w! This hysteresis is both bizarre and common; most people are familiar with the fact that for a single position of the handle, water flows differently out of a faucet depending on whether you have turned it on slowly from the "off" position or slowly decreased the pressure from the "full on" position (see fig. 9).

The important point is this: between w_b and w_c there are two possible states of the system. Knowing the exact equations of motion, including the exact value of the driving frequency, does not suffice to understand the behavior of the system.[10] For if the system is being driven at frequency w between w_b and w_c, its response to the driving force depends on its history. If the system "came from below" (from a lower frequency), it will have a relatively large amplitude of oscilla-

10. Determinism, as uniqueness of evolution, fails here.

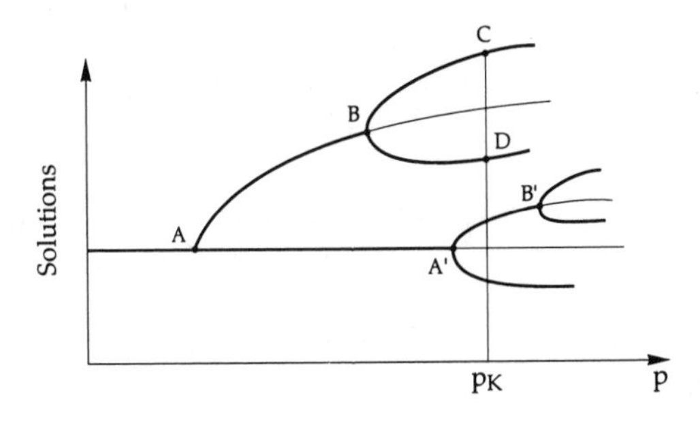

Fig. 10. A bifurcation diagram after Prigogine 1980. From *From Being to Becoming*, by Ilya Prigogine. Copyright © 1980 by W. H. Freeman and Company. Reprinted with permission.

tion, while if it "came from above" (from a higher frequency), it will have a relatively small amplitude.

For systems with hysteresis effects, we cannot understand the behavior of the system without knowing its history, where "history" is understood in the very limited sense of "record of past behavior."[11] In fact, almost any system which undergoes bifurcations as a control parameter is altered is a likely candidate for this type of effect. Prigogine illustrates this point with a bifurcation diagram (see fig. 10).

Suppose we know that the system is in state C after increasing parameter p. "Interpretation of state C implies a knowledge of the history of the system, which had to go through bifurcation points A and B" (Prigogine 1980, 106). In other words, knowing the equations of the system and the value of the parameter p does not suffice to tell us the state of the system, because another point (e.g., D) is also a valid solution there. And furthermore, we cannot understand *why* the system is at point C without understanding *how* it came to be there. Saying "the system is at C because parameter p was increased" will not suffice; reference must be made to which bifurcation path the system followed at B.

11. Dyke 1990 provides a valuable discussion of nonlinear dynamics and historical understanding.

Wesley Salmon has stated, "whether history is classified as a science or not, there can be no doubt that some sciences have essential historical aspects. Cosmology, geology, and evolutionary biology come immediately to mind" (Salmon 1989, 25). Physics may not come to mind because hysteresis effects and bifurcation behavior were typically confined to the ghetto of curious phenomena encountered only in phase transitions like magnetization. (The neglect of hysteresis is allied to a general interest in determinism, and will be more fully discussed in chap. 5.) Chaos theory shows us that the need for diachronic methods of understanding is much broader than previously thought.

The Character of Understanding

We have discussed the object of understanding for chaos theory and its method for addressing that object. It remains to be seen how simple models give us an understanding of the appearance of unpredictable behavior. This last question can be seen as asking what these models do for us that counts as providing understanding. I will address this question by showing that the character of the understanding provided by chaos theory differs from three prominent philosophical accounts of scientific understanding. Specifically, chaos theory does not provide predictions of quantitative detail but of qualitative features; it does not reveal hidden causal processes but displays geometric mechanisms; and it does not yield law-like necessity but reveals patterns.

These three points are directed at three main conceptions of explanation: the epistemic, the ontic, and the modal. These three conceptions are organized around different notions of what it is about scientific explanations which enables them to expand our understanding.[12] The epistemic conception holds that science advances understanding by making events less

12. See Salmon 1989, 130.

surprising, by making more phenomena *expectable*.[13] The ontic conception views progress in scientific understanding as resulting from the disclosure of the hidden causal processes which are responsible for apparently mysterious behavior. And the modal conception sees scientific understanding expanding every time we are able to characterize more events as happening out of necessity.

The Epistemic Conception: Quantitative versus Qualitative Predictability

Some may argue that hardly anyone nowadays thinks that being able to predict the behavior of a system is the same as understanding it. But consider the sentiment expressed by Carver A. Mead, a brain scientist, in the *Chronicle of Higher Education* (October 26, 1988): "If you can properly predict the outcome, there is nothing to learn." And Wesley Salmon characterizes one of the main philosophical conceptions of scientific explanation—what he labels the epistemic conception—as the view that explanations "show that the event to be explained *was to be expected*" (Salmon 1989, 119). Of course expectability is not the same as predictability, but my point is just that one prominent philosophical view would have us clarify the understanding that chaos theory gives us chiefly in terms of what it allows us to say about the behavior of a system before we actually observe it.

The connection between understanding and prediction was made most forcibly in Hempel's deductive-nomological model of explanation. A crucial aspect of this account was the explanation-prediction symmetry thesis, which states that every scientific explanation has the form of a prediction, and that every prediction is also an explanation. It is not important to

13. Salmon treats expectability as only one possible variant of the epistemic conception. Another prominent account that falls under the epistemic conception, the view of explanation as unification, will have to be the subject of later work on the philosophy of chaos theory.

spell out the details of this model and its conceptions of explanation and prediction; the crucial idea is that scientific understanding is gained by constructing explanations of events in the natural world, where an explanation is conceived as an argument showing that, given our knowledge of the laws of nature, "the event in question was predictable had the explanatory facts been available early enough" (Salmon 1989, 48).

The difficulties of this account have generated a great deal of philosophical discussion, but I wish to focus on the central notion that understanding the physical world is the same as being able to predict what happens in the physical world.[14] This notion of scientific understanding runs aground in both directions, as it were. First, it is not the case that we necessarily understand when we can predict. One celebrated counterexample should suffice: the case of ancient Babylonian predictions of lunar eclipses by means of sheer skill at observation and calculation.

> The Babylonians acquired great forecasting-power, but they conspicuously lacked understanding. To discover that events of a certain kind are predictable—even to develop effective techniques for forecasting them—is evidently quite different from having an adequate theory about them, through which they can be understood. (Toulmin 1961, 30)

Second, it is not the case that we necessarily can predict the behavior of a system even though we understand it. For we understand certain chaotic systems (as stipulated above) and

14. The confusion of prediction and understanding may be related to the mistaken identification of all objects of understanding as events. Consider that science is often concerned with processes and not only the events that result from them: "The most spectacular use of mathematics, and especially dynamics, has been to predict accurately and successfully observations of planets and discoveries of new ones, atomic explosions, landings on the moon, and other events. But much of science concerns not prediction so much as understanding—how do galaxies form, how did species arise, how do economies develop?" (Hirsch 1984, 11). See also Roqué 1988.

yet it is impossible to predict their behavior in detail. The physicist Robert Shaw has made this point even more forcefully, by arguing that even if we had a perfect model of a chaotic system, one that exactly accounted for every physically relevant feature, we would still be unable to predict in detail the future states of the actual system. The series of measurements predicted by the model and the series produced by the system itself would have to diverge at a specifiable exponential rate (Shaw 1984, 229).

So chaotic dynamics makes detailed aspects of some systems' behavior unpredictable. This observation can do nothing more than put another nail in the coffin of the long-dead idea that understanding is identical to predictability. After all, the atomic theory was not considered to make the world incomprehensible just because it rendered predictions monstrously difficult due to the number of entities involved. Nor did the adoption of quantum mechanics seem like a step backward in our understanding of the physical world because it rendered some predictions impossible.[15] On the contrary, these theories opened up new areas to fruitful investigation and insights. Atomic theory made it possible to find and account for general relationships among the large-scale or average properties of matter, and quantum mechanics let us know with great accuracy just when, and why, the impossible-to-predict behavior was to be expected.

Chaos theory, too, does not lessen our understanding or render much of nature incomprehensible. For in the first place, it gives us new general information about the relationships between the large-scale properties and long-term behavior of systems, even allowing new predictions. And in the second place, like quantum mechanics, it gives an intelligible and enlightening account of when predictability will go out

15. Of course there are still philosophical questions about whether quantum mechanics provides genuine explanations for physical events, but these problems do not arise from the fact that certain predictions are ruled out by the mathematical formalism.

the window, and even an account of how it is that this happens.

These considerations lead us to think that the conception of understanding as predictability contains some truth. For two aspects of the understanding gained from chaos theory are just these: it allows for new predictions at a different level of detail, and it provides an account of the limited predictability of chaotic systems. I suggest that although predicting is not the same as understanding, we are justified in thinking that this enabling of new and interesting predictions and accounting for the limits of predictability are *signs* of understanding.

It may seem odd to assert that chaos theory has "great predictive power," but such claims appear in the scientific literature and are not meant to be tongue-in-cheek (see for instance Bergé, Pomeau, and Vidal 1984, 265). These assertions refer not to an ability to predict the exact value of some property of a system, but to an ability to foresee and understand changes in the overall behavior of that system. A scientist would not claim to expect some particular value for, say, the velocity of a fluid at some particular point at some particular future time. But he or she may well claim that as you increase the speed of the fluid flow, one would expect the flow to become turbulent at some particular speed which could be determined in advance (Ruelle and Takens, 1971, 168).

We find some of the best examples of this kind of "higher level" prediction in systems which undergo the period-doubling route to chaos. For these physical systems, which include fluid convections, forced electrical oscillations, and the development of instabilities in lasers, one can make important predictions (Collet and Eckmann 1980, 25).[16] For instance, if we observe two successive period-doubling bifurcations in the system at two values of a control parameter p, we can predict where other bifurcations will occur, when chaotic behavior is

16. For a useful presentation of many actual experimental situations that display the period-doubling route to chaos, see Moon 1987.

expected to begin, and when it will yield to a new periodic regime (Collet and Eckmann 1980, 51). The more measurements we make, the more accurate these predictions will be.

The crucial point here is the distinction between specific *quantitative* predictions, the usual sort of which are impossible for chaotic systems, and *qualitative* predictions, which are at the heart of dynamical systems theory. Quantitative investigations can provide very accurate information about a dynamical system by solving the equations of motion, but for nonlinear systems this information is typically limited to just one solution and some small vicinity around it, and any accuracy secured rapidly disappears with time. Qualitative understanding is complementary; it predicts properties of a system that will remain valid for very long times and usually for all future time. It gives "the general informations and the great classifications," by dealing with questions such as the periodicity and stability of orbits, the symmetries and asymptotic properties of behavior, and "the structure of the set of solutions" (Marchal 1988, 5; see also Abraham and Shaw 1982, 27).

As Poincaré wrote,

> In the past an equation was only considered to be solved when one had expressed the solution with the aid of a finite number of known functions; but this is hardly possible one time in a hundred. What we can always do, or rather what we should always try to do, is to solve the qualitative problem so to speak, that is to try to find the general form of the curve representing the unknown function. (Quoted in Hirsch 1984, 19)

In chapter 5 I will take up the question of why it took so long for researchers to take this advice. For now it will suffice to note that nonlinear dynamical systems theory addresses itself to the task of providing precisely this qualitative type of understanding. As we have seen in chapter 1 and in "The Object of Understanding" above, chaos theory seeks to learn when a system manifests periodic behavior, when this "form of curve" becomes unstable, and so forth. In general, chaos theory often seeks to understand the behavior of a system by reconstructing its attractor, and knowing this attractor gives us qualitative understanding. In the terminology of the semantic view, chaos theory includes theoretical hypotheses that assert relationships of qualitative (or topological) similarity between its abstract models and the actual systems it studies. "Dynamics is used more as a source of qualitative insight than for making quantitative predictions. Its great value is its adaptability for constructing models of natural systems, which models can then be varied and analyzed comparatively easily" (Hirsch 1984, 11).

To further drive home the point about experimentalism made earlier, consider this connection between the noninferential method of chaos theory and its qualitative character. In seeking to understand the transition to chaos, Leo Kadanoff describes a two-stage procedure. The first stage is qualitative analysis, wherein one "mainly uses computer experiments to gain a description of the behavior" (Kadanoff 1985, 60). For instance, a numerical simulation of the behavior of the system may reveal a period-doubling cascade. One should then obtain the scaling relations by examining the sequence of bifurcation points, identify the relevant symmetries of the situation, and test the generality of these features by exploring other, similar maps.

"Once this qualitative exploration is completed," he writes, "publish" (p. 61). The second stage of the analysis, the implementation of a renormalization group following Feigenbaum's example, will rigorously establish the universality of the observed qualitative results. But Kadanoff speaks with something like disdain for this step: "the return on effort invested is not high." It is all well and good to be able to prove universality, and one may obtain a slightly more accurate value for the scaling exponents, but the actual proof will not usually produce "new insights." Now perhaps this disdain is caused more by the traditional differences in approach between mathematicians and physicists than by anything fundamental to the character of chaos theory. Nonetheless, the understanding of the situation is of a qualitative nature, and it is manifestly gained by numerical experimentation and not by rigorous schemes of inference.

So chaos, to make the analogy with atomic theory, puts certain detailed quantitative predictions out of reach while enabling us to make quite useful predictions about the qualitative features of the systems involved. And just as reasoning about the behavior of atoms can yield information about the average properties of a collection of atoms (as in statistical mechanics), chaos theory also provides statistical information. For although a chaotic system will display behavior that travels randomly through a range of allowable states, an analysis of the bifurcation diagram or the attractor can provide information about how much time the system will spend in one region or another. "Since an analytical description of the chaotic evolution of individual initial conditions is impossible, the best we can hope for is a statistical theory which predicts the likelihood of the variable x_n taking on any particular value" (Jensen 1987, 172).

Such a statistical theory can have several uses. First, once we find the probability distribution for a chaotic system, "one can calculate the mean square amplitude, mean zero crossing times, and probabilities of displacements, voltages or stresses exceeding some critical value" (Moon 1987, 153). For example, from the most general features of the vibratory behavior of a steel beam we may be able to predict when it would fracture under stress, although such predictive ability remains to be developed.[17]

Perhaps more important is that such a statistical theory can tell us the "range of validity" for its generalizations (Jensen

17. Another use for such a statistical theory is to generate probabilistic generalizations from which to deduce the probabilities for a chaotic system to take on some particular state. When the system then took on this state, we could (following Peter Railton's "Deductive-Nomological-Probabilistic" model of explanation) explain why such an event occurred (Railton 1981). I will leave a detailed working-out of this project for another time.

1987, 172). That is, chaos theory can provide an account for just when and to what extent its generalizations are legitimate. Here we see a perfect example of the way chaos theory addresses its object of understanding: it constructs simple models that yield qualitative theories and can account for their range of validity.

The Ontic Conception: Causal versus Geometrical Mechanisms

The ontic conception holds that our understanding of the world is increased "when we obtain knowledge of the hidden mechanisms, causal or other, that produce the phenomena we seek to explain" (Salmon 1989, 135). These hidden mechanisms are usually characterized as causal processes or interactions that transmit influences. In this section I will argue that chaos theory does indeed give us understanding by showing us the mechanisms responsible for unpredictable behavior, but that these are not causal processes. Instead, these mechanisms are best characterized as geometrical, and chaos theory tells us how they function. So I am arguing here only against the view that it is always appropriate to see understanding as providing knowledge of underlying causes. In the next section, I will argue that these mechanisms are not "lawful" mechanisms, either.

James Woodward has already presented a persuasive account of the limitations of a strictly causal mechanical theory of explanation. He cites examples of physical systems with many interacting processes where "it is often hopeless to try to understand the behavior of the whole system by tracing each individual process. Instead, one needs to find a way of representing what the system does on the whole or on the average, which abstracts from such specific causal detail" (Woodward 1989, 362–63). The reader will recognize here the nonreductive methodology typical of chaos theory. What is more striking is that the tracing of individual causal processes is impossible for chaotic systems not because of the large number of interacting effects. Instead, intractably complicated chaotic behavior can appear even in mathematically simple systems.

Woodward contends that biology, psychology, and the social sciences commonly proceed by ignoring the level of "fine-grained, microreductive causal detail" (p. 365). Researchers in these fields are more interested in finding "common patterns or regularities at a more macroscopic level of analysis." This notion of pattern-finding will reappear in the next section. But note that this procedure of grouping together systems with different underlying causal substrata in order to study their qualitative behavior is central to the method of chaos theory. In other words, an account of scientific understanding as the disclosure of hidden causal processes is not only inadequate for the biological and social sciences, as Woodward rightly claims, but it is inadequate for the physics of nonlinear systems as well.

As Bergé, Pomeau, and Vidal write in the introduction to their survey of chaos theory, the method of studying dynamical systems treats them without any reference to "the actual matter through which they are manifested" (p. xiii). This modeling procedure ensures the generality of the results and "it gives us indications about a physical system *without knowing specific details*" (Collet and Eckmann 1980, 25, emphasis in the original). Chaos theory does not reveal the causal process at work because it does not need to.

Furthermore, it is impossible in principle to trace out the workings of the actual causal mechanism in a chaotic system. To do this would require calculations using the evolution equations of the system. As discussed in chapter 2, even if one is in possession of the exact equations of motion, the slightest degree of inaccuracy in numerical calculation or in specifying initial conditions will rapidly lead to predictive hopelessness. As one researcher puts it, after a certain specifiable time "the initial and final states will be causally disconnected, certainly from the point of view of the observer" (Shaw 1984, 220). This statement should not be read as an invocation of mysterious acausal influences. Instead, it means that it is "transcendentally" impossible to trace the actual causal influences that lead from one state to a later one. Not even an "ideal explanatory text" could contain the full causal account, unless the sense of idealization involved forfeits any claim to being possible in principle.

There is some indication that Salmon has changed his position on the centrality of causal mechanisms between his 1984 book and his comprehensive 1989 review article. In the former it seems as though he defended the notion that scientific explanation always works by revealing underlying causal processes, while in the latter he expands this idea to include "other" types of mechanisms such as statistical or structural laws as well (see Salmon 1989, 121). Such an expanded view would not find itself contradicted by nonlinear science, for while chaos theory does not pay any attention to causal mechanisms, it does concentrate on revealing mechanisms of another sort: geometric mechanisms.

What could it possibly mean to say that chaos theory gives us understanding by revealing the geometric mechanisms responsible for unpredictable behavior? Consider the following account given by Albert Libchaber:

> A physicist would ask me, How does this atom come here and stick there? And what is the sensitivity to the surface? And can you write the Hamiltonian of the system? And if I tell him, I don't care, what interests me is this shape, the mathematics of the shape and the evolution, the bifurcation from this shape to that shape to this shape, he will tell me, that's not physics, you are doing mathematics. . . . Yes, of course, I am doing mathematics. But it is relevant to what is around us. That is nature, too. (Quoted in Gleick 1987, 210–11)

Chaos theory gives us the "geometry of behavior," to use the phrase that appears in the title of Abraham and Shaw's superb graphical introduction to dynamical systems theory (1982). That work is dedicated to fostering understanding by actually showing, in four-color pictures, the qualitative behavior of dynamical systems and how that behavior changes as parameters are varied. The authors are not content to state that a result has been rigorously proven; they insist on displaying the geometric features responsible (see, for instance, their discussion of Peixoto's theorem, pp. 175–79).

Information about the geometrical features of the model's orbits tells us about the qualitative features of the physical system's behavior: "if we can understand the topology and stability" of the attractor for the Lorenz equations, for example, "then we will know the long-term behavior of all trajectories" (Sparrow 1986, 113). The theoretical hypotheses of chaos theory assert relationships of topological similarity, not congruence of physical causes, between its exemplary models and actual systems.

Recall the discussion of strange attractors in dissipative systems in chapter 1. The abstract geometric mechanism of stretching and folding accounts for the sensitive dependence on initial conditions responsible for unpredictability as well as for the fractal structure that persists in the presence of randomness (Bergé, Pomeau, and Vidal 1984, 205). Consider that Ruelle and Takens presented their groundbreaking account of the transition to chaos under the heading "A Mathematical Mechanism for Turbulence" (1971, 171). One final example, from the theory of conservative systems, will drive home the fact that chaos theory provides understanding by displaying the geometric mechanisms responsible for unpredictable behavior.

The central result in the study of Hamiltonian dynamical systems, where total energy is conserved, is the celebrated Kolmogoroff-Arnold-Moser theorem, or KAM. Briefly stated, this theorem provides that if you start with a "nice," strictly linear, fully integrable system and then add a small nonlinear perturbation, the system will remain qualitatively similar. An integrable system, even if there is no exact analytic solution to the equations of motion, will nonetheless be confined to the surface of a torus in state space. In the simple case of two

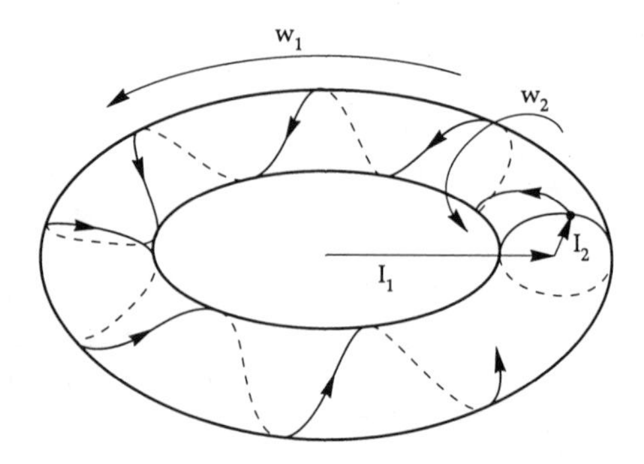

Fig. 11. KAM torus

integrals of the motion, the orbits are confined to a two-dimensional torus (see fig. 11).

The point representing the state of the system (the state point) moves on the surface of the torus, circling the "hole" of radius I_1 in the donut-shaped torus with angular frequency w_1 and going around the smaller radius I_2 with frequency w_2. If w_1/w_2 is a rational number, the motion is periodic and the state point will eventually return to exactly the same spot. If w_1/w_2 is irrational, the motion is quasiperiodic and the orbit will wind densely around the torus, eventually coming as close as you want to every point on it. The KAM theorem states: most sufficiently irrational tori survive a small perturbation. The strictly periodic orbits rapidly become unstable: points that begin near them wander off. But if w_1/w_2 is highly irrational, that is, far from a rational number p/q with p and q both small integers, then the quasiperiodic orbit will not be greatly deformed by introducing, say, a small nonlinear term.

As long as there are many tori that have survived without drastic deformation, the motion of a trajectory from an unstable orbit will still be confined between two "donuts," one closely nested inside the other. Such behavior is still, for practical purposes, integrable (Chirikov 1979, 310). As the nonlinear perturbation increases, however, more and more tori are "destroyed" and their trajectories are free to wander over much of the allowable state space region. Eventually, most of the tori are radically deformed and the state space is littered with the now-unstable periodic orbits.

This situation brings about chaotic behavior in the following way: as the perturbation is increased further the realm of the destroyed tori increases, so that their orbits will roam over more and more of the allowable state space region.[18] And that region is now littered with the remnants of periodic orbits that have become unstable and wildly convoluted. As a state point wanders about it will soon come very close to one of these orbits. Of course the periodic orbit will not "capture" the point because it is unstable—the point must soon wander away from it. But in the meantime the point will follow it around for a short while and then come across another periodic orbit that has gone unstable and wander with it for a while. The whole region is thick with unstable orbits and the point swings from one to another, circling for a while and then reeling off.

Now imagine yourself in the place of the state point. Imagine you are square dancing and you swing your partner a few times, then your corner, then the square breaks up and the whole room is filled with partners who swing you a few times, no way to tell how many, but their hands are slippery so you fly away and then suddenly there is another person in front of you and there is just time for you to swing with them for a while. There are so many potential partners that it is guaranteed that soon another one will be in the right place at the right time to whisk you away. Each is twirling at a different rate and each has slippery hands, so you will whirl about from place to place erratically. Thus, chaos results from a superabundance of order: too many opportunities for periodic behavior, all unstable.

Note that for the logistic map with $a = 4.0$, the unit interval is filled with an infinity of unstable periodic orbits of lit-

18. This account relies heavily on Walker and Ford 1969, Jensen 1987, and Chirikov 1979. The square-dancing metaphor is new.

erally *every* period, so you will always be near one, falling into its groove for a while but eventually being forced to leave. This fact explains why some definitions of chaos require dense periodic orbits (see Devaney 1986, 50). Here is an example of a geometric mechanism: chaos happens because of or through the dense packing of unstable periodic orbits.

This fanciful account is a geometric story, a moving picture of what is going on in abstract dynamical state space. It is constructed with the help of both rigorous mathematical theorems and numerical simulations that show what is going on to account for the results of those theorems. Chirikov described his investigation in just such a way: "one may try to investigate qualitatively, I would say even graphically, particular mechanisms responsible for the destruction of integrals of motion under a sufficiently strong perturbation" (Chirikov 1979, 285). Thus, he used numerical experiments to investigate the qualitative features of dynamical systems and discovered a new criterion for when to expect the onset of chaotic behavior. Chirikov's investigation provided new understanding by displaying a geometric, not a causal, mechanism.

The Modal Conception: Law versus Order

Adherents of the modal conception of explanation see a lack of understanding wherever we must admit that an event "just happens." Science, according to this account, furnishes understanding by widening the scope of events that happen out of necessity and not contingency. By revealing laws of nature and subsuming phenomena under them, science makes the world comprehensible. As Nicholas Rescher states:

> A recourse to laws is indispensable for scientific explanation, because this feature of nomic necessity makes it possible for scientific explanations to achieve their task of showing not just *what* is the case, but *why* it is the case. This is achieved by deploying laws to narrow the range of possible alternatives so as to show that the

> fact to be explained "had to" be as it is, in an appropriate sense of this term. (Rescher 1970, 13–14, quoted in Salmon 1989, 92)

If science gives us understanding of the physical world by expanding the scope of nomic necessity, then chaos theory does not give us any understanding at all. Nomic necessity requires that from universal laws and statements of initial conditions we can generate with deductive rigor the uniquely determined past and future behavior of a system in fine detail. But chaos theory is neither strictly deductive, nor quantitatively predictive, nor globally deterministic. And furthermore, researchers in chaos theory do not portray their work as discovering new laws of nature.

Now of course chaos theory establishes powerful generalizations, such as the universal scaling relations for systems that undergo period-doubling. But there is as yet no way to tell beforehand, solely on the basis of equations of motion, which route a system will take to chaos (Bergé, Pomeau, and Vidal 1984, 265). That is, given a physical system and the full set of equations governing its behavior, there is no set of necessary and sufficient conditions that would allow us to foresee which type of transition to unpredictability it will follow. We know that *if* the system exhibits period-doubling bifurcations then it will obey the scaling relations, and this allows us to infer that the behavior is governed by a map with a quadratic extremum. But there simply is no "universally valid law from which the overall behavior of the system can be deduced. Each system is a separate case; each set of chemical reactions must be investigated and may well produce a qualitatively different behavior" (Prigogine and Stengers 1984, 144–45).

The modal conception appeals to our sense that we understand more when we can take puzzling or mysterious or surprising aspects of the natural world and fit them into what we already know. The previously incomprehensible aspect is now seen to be perfectly "normal," since it did not "just happen." Rather, it "had to" be that way—even if we cannot in practice trace out the sufficient reasons for the event, we can nonetheless see that it fits into an intelligible pattern.

Chaos theory takes up and preserves this emphasis on finding patterns and connections, while jettisoning the requirement that the patterns must yield necessity in a detailed and deterministic sense. Researchers may be fascinated by the development of chaotic motion, but "at the same time they look for regularities" (Haken 1981, 7), using computer graphics to "identify and explore ordered patterns which would otherwise be buried in reams of computer output" (Jensen 1987, 168).

Some may contend that this search for patterns actually strives to discover new laws governing qualitative features of systems.[19] But in the face of the nonreductionist, nondeductivist, diachronic methodology of chaos theory, it does violence to the actual practice of this science to force it into the mold of a law-seeking activity. Moreover, the conception of understanding as the discovery of laws has deep connections to the doctrine of determinism as total predictability. Far better to consider chaos theory as a search for *order,* a concept broader than law.

Evelyn Fox Keller describes the emphasis on law in science as an inappropriate limitation on our notion of scientific understanding. Granted that some laws of nature are statistical or phenomenological, still "in many, if not most, scientific disciplines the finality of a theory continues to be measured by its resemblance to the classical laws of physics" (Keller 1985, 133). Those laws promise a deterministic universe of unique evolution and fuel the methodological goal of total predictability. The law-based conception of understanding seeks iron-clad rules that will dictate why things are constrained to turn out the way they do. Such an approach would typically respond to chaotic behavior dismissively, assigning it to un-

19. Perhaps such laws would deal with topological properties, which remain invariant up to a homeomorphism, while failing to preserve metric properties. On the subject of the plurality of interesting questions about a system, see Wittgenstein 1958, 179.

controlled outside causal influences ("noise") or writing it off as the unintelligible result of too many competing and interacting simple mechanisms (the Landau model).

But chaos theory looks to the geometric mechanisms that will show how patterns arise alongside unpredictable behavior, providing an understanding of "how it happens" rather than of "why it had to happen." Such an investigation reveals order:

> Order is a category comprising patterns of organization that can be spontaneous, self-generated, or externally imposed; it is a larger category than law precisely to the extent that law implies external constraint. Conversely, the kinds of order generated or generable by law comprise only a subset of a larger category of observable or apprehensible regularities, rhythms, and patterns. (Keller 1985, 132)

Keller recommends an emphasis on order rather than law so as to revise our conceptions of science and the natural world for the better. In many ways, the success of chaos theory validates her suggestions. As evidence for such a claim, consider her statement (Keller 1985, 134) that an interest in order rather than law may be expected to lead to a shift toward "more global and interactive models of complex dynamic systems." And likewise, consider her description (page 135) of the way the law-based conception runs up against the limits of its appropriateness, necessitating "the development of new mathematical techniques" that are "better suited to describing the emergence of particular kinds of order from the varieties of order that the internal dynamics of the system can generate."

One can also use Keller's account of science as the search for order to make better sense of the character of the understanding that chaos theory provides. In this spirit, I contend that it is better to see chaos theory as providing insights into order than it is to try to fit it into a model of science as a search for laws. The theoretical hypotheses of chaos theory assert that certain abstract models and certain actual systems are instances of similar varieties of order.

Indeed, the concept of order seems to sum up the answer to the question, What kind of understanding does chaos theory provide? The object of this understanding is the way in which unpredictable behavior and patterns come to appear. The method of understanding their appearance is by the construction of models, not by breaking systems into their components and then constructing ahistorical deductive schemes, but rather by using experimental procedures that concentrate on holistic properties and historical development. And the character of the understanding these models provide is that of qualitative expectability, geometric mechanisms, and order. That is, these models count as providing understanding because they show us how chaos happens. Here is a sense in which we find order in chaos.

Dynamic Understanding

By way of summary, I propose that the kind of understanding provided by chaos theory be called "dynamic understanding." This label conveniently makes us aware of three senses of the word "dynamic." First, it calls to mind the connection with dynamical systems theory, the qualitative study of the behavior of simple mathematical systems. Second, it connotes change and process, tying together the various uses of the word "how," which have permeated this chapter. Chaos theory lets us understand how patterns and unpredictability arise by showing us how certain geometric mechanisms bring them forth. And finally, "dynamic understanding" is parallel to Keller's notion of "dynamic objectivity," which characterizes scientific investigations that do justice to the complexity and order in their objects of study.

So chaos theory provides us with understanding that is holistic, historical, and qualitative, eschewing deductive systems and causal mechanisms and laws. These conclusions seem to support the idea that nonlinear dynamics shares much in common with the nonscientific or non-Western intellectual traditions mentioned earlier. Invocations of these commonalities sometimes make it sound as if chaos theory stands in drastic contrast to the purportedly heartless calculation of traditional science. The time has come to evaluate the possible claim that chaos theory represents a radically new or culturally superior form of science.

Nonlinear dynamical systems theory is holistic to the extent that it studies properties of physical behavior that are inaccessible to microreductive analytical techniques. But it nonetheless proceeds by massively simplifying the models it studies and carefully isolating experimental setups. It often pays more attention to numerical simulations than to deductive structures, but there is no place in the journals where the relaxation in mathematical rigor is construed as an excuse for sloppy data collection or vague appeals to intuition. Effects like hysteresis abound, but there is no sign that physical science is about to accept a compelling historical narrative as an adequate account for, say, aperiodic behavior in lasers.

The character of the understanding we get from chaos theory is qualitative, yes, but this must not be construed as if "qualitative" is to be contrasted with "mathematical." The dynamical systems approach gives no indication of reversing the process, begun by Galileo, of the ever expanding mathematization of the world. Far from creating a space for the reappearance of qualitative properties in the sense of subjective, sensuous experiences, chaos theory strives to apply mathematical techniques to phenomena like turbulence that were once a repository for Romantic notions of sublime Nature resisting the onslaught of human rationality.

To see chaos theory as a revolutionary new science that is radically discontinuous with the Western tradition of objectifying and controlling nature falsifies both the character of chaos theory and the history of science. A new science such as chaos theory can reveal the limits of standard methodological approaches to understanding the world and impel a reconsideration of the metaphysical views that undergirded them. But any expectation that chaos theory will re-enchant the world will meet with disappointment.

Perhaps research into nonlinear systems does provide a new kind of understanding, which could lead to a new conception of nature that accepts randomness and contingency. But why, then, do scientists engaged in studying chaotic systems still describe chaotic behavior as "pathological"?[20] Why is there so much talk of the need to cope with, or control, or even "cure" chaotic systems?[21] Could it be that these researchers themselves have not learned the lessons of the beauty and diversity of chaotic behavior? Some researchers persist in "settling for" chaos theory as "all we can get" in certain unfortunate situations whose ubiquity seems like a conspiracy to rob us of neat Newtonian systems.

The apparent reluctance of many scientists to embrace wholeheartedly the new form of understanding provided by chaos theory raises anew the concern voiced at the beginning of this chapter: perhaps chaos theory does not provide us with any "real" understanding at all. After all, chaos theory does not introduce any fundamental revisions in our laws of nature as quantum mechanics or relativity did; it merely introduces new stop-gap measures like qualitative analysis. Perhaps it is not yet really a Science but only the beginnings of one, struggling to make interesting observations or empirical generalizations like Brahe or Kepler, but still awaiting the formalization that would mark its legitimacy.

Such suggestions invite four responses. First, philosophical discussions will suffer if they decide beforehand what can or

20. See, for example, Jensen 1987.

21. See Bergé, Pomeau, and Vidal 1984, 265; Moon 1987, 7; and Jensen 1987, 169. Consider also James Yorke, one of the founding fathers of chaos theory, who refers to the need to know about chaos as analogous to the need of an auto mechanic to know about "sludge in valves" (quoted in Gleick 1987, 68).

should count as scientific understanding. The best basis for saying to scientists, "No, you are mistaken, you only think you understand" is an internal critique: a demonstration that the methods in question do not satisfy the scientists' own conception of understanding. As I have argued, debate is most profitably engaged by first examining claims about the nature and scope of the purported understanding, and then proceeding to a possible challenge to the current conception of what understanding can or should be. Second, different parts of the world have different types of understanding appropriate to them. Chaos theory uses the appropriate methods to get an appropriate understanding of certain types of systems. It is only because science focused for so long on integrable, linear systems that chaos theory could ever be seen as second-rate, "make-do" understanding. Furthermore, we should be aware that linear systems are very rare and periodic behavior can be very boring and even fatal.[22]

Third, it may be more responsible to leave it up to future developments to judge. The only claim being made now is that we have *some* understanding and that it is not of certain other types. Perhaps it will later be seen as only a small amount of tentative understanding, but it is still worthwhile to investigate its nature. And finally, a major revision in scientific methodology should not be disparaged on the grounds that fundamental theoretical structures are left unchanged. Chaos theory does not challenge any of the most basic physical laws: that has been stipulated from its beginning. But recall the central feature of the semantic view: models, not laws, form the heart of a science. Chaos theory insists that we rethink all our notions about the applicability of classical models.

These responses deal with the philosophical challenge that would claim that chaos theory does not represent a new type

22. I have in mind here recent investigations that indicate that aperiodic or chaotic dynamics are typical of the healthy functioning of the human neurological and metabolic systems. See Glass and Mackey 1988; Prigogine and Stengers 1984, 134 and 153; Gleick 1987, 239; and Shaw 1984.

of understanding. But what of the persistent attitudes of some scientists, noted above, who seem not to have embraced this new understanding with full enthusiasm? Their case will be addressed in chapter 5, in the context of the question, Why did it take so long for chaos theory to be developed?

5

Beyond the Clockwork Hegemony

> The truth is that some phenomena are regular and some are not. Western Science selects as its subject matter those that are regular and then finds that it can predict their behavior. But this is no basis for concluding that irregular and irrational phenomena are not important or are trivial.
>
> —Nwankwo Ezeabasili (1977)

A Historical Question

Chaos theory as a scientific research project is no more than thirty years old, and interest in nonlinear dynamics continues to grow. From a historical perspective, one may ask what factors account for this explosion of interest. Why has chaos theory become the object of so much interest and even hype? The answer may lie in some feature of current intellectual development in the sciences or in some aspect of contemporary society or (more probably) in some complex combination of several influences.

This chapter addresses a different question about the historical development of chaos theory, however. Instead of asking why chaos theory is so successful now, I would like to examine why it took so long for scientists to focus attention on these phenomena. After all, Poincaré and Birkhoff discussed nonlinear dynamical systems with sensitive dependence on initial conditions near the turn of the century; yet their mathematical discoveries waited half a century before their fruitfulness for physical science was explored. If chaotic behavior is all around us, as scientists now tell us, and if the mathematics to study it was available seventy-five years ago, why did it take so long for chaos theory to develop?

The most common answer to this question maintains that chaos theory relies heavily on digital computers: there was no way to study chaotic behavior scientifically until these computers were invented. I contend that this is only a partial answer, because alternative computational resources were available. A fully adequate account of the historical development of chaos theory must include discussion of broader cultural factors. One such factor I will consider is the social interest in the exploitation of nature, an interest that contributed to the institutionalized disregard of physical systems not readily amenable to analysis and manipulation.

I will proceed by first documenting the situation I call "the nontreatment of chaos," making a case that this nontreatment raises a genuinely significant historical question. Next I will evaluate two explanations for this situation that have appeared in the scientific literature: the necessity of computers mentioned above and the institutional emphasis on linear systems. After arguing that these answers are insufficient, alone or together, to account for the situation, I will examine some of the broader cultural factors that form part of a more satisfactory answer to this question.

Without the resources or the expertise to approach this question as a historian or a sociologist of science, my account must be seen as a sketch for such a project. As part of this sketch, I will indicate where empirical evidence of a historical or sociological nature could bolster or weaken my argument. My primary goal, however, will be to construct a case for the plausibility of the following philosophical proposition: cultural biases can profoundly affect the historical development of physics by influencing the scientific community's notions of what counts as an interesting or worthwhile scientific phenomenon.[1] To this extent, the development of chaos theory

1. This chapter will focus on the field of physics and especially the historical development of the study of oscillations in nonlinear systems. But

provides evidence in support of the contention of feminist philosophers that science, *even* physics, is significantly influenced by cultural factors such as ideology.

The Nontreatment of Chaos: Mathematical Precursors

Two central aspects of chaos theory are the mathematical study of abstract dynamical systems and the application of these dynamical models to complex behavior in actual experimental systems. The nontreatment of chaos consists in just this: both the mathematical models and the experimental systems lay unaddressed in the published literature for well over fifty years. Much of the relevant mathematics was available but not utilized. Many of the relevant experiments were feasible but not carried out. Moreover, when one of these experiments was in fact conducted, the chaotic behavior was dismissed and subsequently ignored.

It is not the case that the prerequisites for chaos theory could have been developed but were not. Instead, these prerequisites were constructed and then neglected, passed over, ignored. The nontreatment of chaos is not a puzzle about why scientists took a long time to find a hidden treasure; it is a puzzle about why they found it and summarily passed it by as dross, continuing to dig elsewhere. In this section I will discuss the mathematical precursors of chaos theory and later take up the early experimental signs of chaos.

Henri Poincaré has posthumously earned the title of the father of the mathematics of chaos theory. His "geometric" approach to the study of dynamical systems provides the standard for current research into nonlinear behavior. He pioneered the use of state space to look at the qualitative features of the dynamics of a system, and the fundamental ideas for bifurcation theory and ergodic theory alike begin with him. Yet "it is notorious that the physicists of most of the twentieth

---

many of the issues discussed are relevant to general features of the history of the physical, biological, and social sciences.

century had little appreciation of Poincaré's work" (Wrightman 1985, 2).

The problem of the ultimate stability of the solar system concerned scientists at least as early as Leibniz. It formed the subject matter of Poincaré's great work on celestial mechanics in 1892, in which he provided a proof of the very weakness that Leibniz suspected in Newton's system of mechanics: there is no way to demonstrate rigorously that the solar system will remain forever in roughly its present configuration. The "many-body problem" does not admit of a simple analytical solution, so we may never know whether the earth will one day crash into the sun. Poincaré's analysis made one major element of the putative grand design of the universe inaccessible to human reason, but it also yielded "the first appearance of a chaotic limit set in the mathematical literature" (Abraham and Shaw 1984, 3). Poincaré found the tangled homoclinic trajectories that yield sensitive dependence on initial conditions in conservative systems, but almost all scientists ignored these results for some fifty years (Jackson 1989, 1; Reichl and Zheng 1987, 20).

Although these aspects of Poincaré's work were passed over by physicists, they fared better with some mathematicians. George David Birkhoff took up the geometric approach and continued the study of instability in the context of celestial mechanics. In the process, "he found the first example of what would nowadays be called a *strange attractor*," which he called a "remarkable curve" (Wrightman 1985, 14). The chaotic attractor thus first appears in Birkhoff's work of 1916, published later in 1932 (Abraham 1985, 117). Still, the relevance of this research to physics remained unexplored, even after Levinson proved in 1948 that the Birkhoff attractor could occur in the behavior of a forced oscillator in three dimensions. Indeed, "if Poincaré has been little appreciated among physicists, Birkhoff's work on dynamical systems has been more or less unknown" (Wrightman 1985, 14).

Beginning with Lyapunov's pioneering investigation of stability, much of the most important work on nonlinear mechanics was carried on in the Soviet Union. Although the celebrated KAM theorem of 1963 occurs right at the birth of chaos theory, much research was done on structural stability and bifurcation theory well before that date. Of particular note is the mid-century research of the Andronov school, which still finds use today within chaos theory. As Francis Moon writes,

> the novitiate to the field of nonlinear dynamics may be misled by the current interest in chaos to conclude that the field lay dormant in the prechaos era. However, a large literature exists describing mathematical perturbation methods for calculating primary and subharmonic resonances, as well as the stability characteristics of solutions to nonlinear systems (Moon 1987, 190).

Finally, the mathematics of chaos involves not only certain analytical tools developed by Poincaré and others but certain mathematical models. Some of these models are quite new, but others were widely used in the early part of the century. Although these models were carefully investigated for certain values of the parameters, the parameter values where chaos appears were never discussed. Robert May points out that the higher parameter values of the celebrated logistic map, which is now one of the central models of chaos theory, lay unexplored in the literature of population biology and economics (May 1976). The equations for the oscillators of Duffing and Van der Pol also display chaotic solutions; these equations were published and even studied, but with the exception of Levinson, noted above, the parameter values that led to chaos were circumnavigated (Moon 1987, 88). The mathematical models for chaos theory, and the tools to study these models, lay unused by scientists for many years.[2]

2. On the preexistence of the mathematics of chaos theory, James Yorke states that Gauss himself studied the dynamics of the doubling map (Yorke 1990), and Christian Mira claims that the quadratic map (equivalent to the logistic map) was studied in full by Myberg as early as 1958, some fifteen years before May's review of the work of the early seventies (Mira 1986, 260).

The Nontreatment of Chaos: Experimental Precursors

Although much of the relevant mathematics existed in the early part of the century, one may still question whether this presents anything more than an example of the fact that scientists do not always make use of the mathematics available. After all, there is much interesting and worthwhile mathematics that has no relevance to any physical situation, and one can hardly expect physicists to search continually through the literature in hopes of finding something promising. The nontreatment of chaos does not require explanation until we see that there were indeed physical phenomena and even experimental situations that called for but did not receive these novel mathematical approaches.

In some sense, of course, chaos has surrounded us since human life began. Natural processes provide examples of systems with sensitive dependence on initial conditions wherever we look, and human-built devices confront us with chaotic behavior in an even more scientifically accessible form. As Francis Moon writes, "engineers have always known about chaos—it was called noise or turbulence and fudge factors or factors of safety were used to design around these apparent random unknowns that seem to crop up in every technical device" (Moon 1987, 6). My claim regarding the experimental precursors of chaos theory is twofold: first, experimental apparatus for studying chaotic behavior was available well before 1975 and, second, chaotic behavior was observed and recorded by researchers well before that date. Scientists "saw" chaos in the sense that their carefully controlled experimental apparatus yielded measurements of complex and unpredictable behavior in simple physical systems. Scientists "ignored" chaos in the sense that these observations were passed over and not considered worthy of further investigation.

May has recently written of population biologists who found chaotic dynamics in their models but did not pursue them (May 1987, 32).

The first of these points is a simple one. The apparatus necessary for carefully observing chaotic behavior need not be exotic or expensive or terribly high-tech. True, Libchaber and Maurer's study of convection in liquid helium (1982) or Gollub and Swinney's study of turbulence between rotating cylinders (1975) utilized technologies that were unavailable to earlier researchers. But the "bouncing rotator" studied by Michael Berry displays the most important features of chaotic behavior and yet it is identical to the "space ball" toy commonly sold in airport gift shops (Berry 1986). And the exemplary work of Robert Shaw utilized an experimental setup that has been readily available for centuries: a dripping faucet (Shaw 1984). The material substrates for current experimental research in chaos—fluid flow, mechanical oscillations, and nonlinear electrical circuits—simply do not present the kind of challenges to engineering or funding that could be solved only recently.

So chaotic systems could have been investigated much earlier. The second point to be noted is that, in fact, they were. Chaos, in the form of intermittency, was "observed by Reynolds in pipe flow preturbulence experiments in 1883" (Moon 1987, 181). Another dramatic example is the report of Van der Pol and Van der Mark, published in 1927, on the observation of frequency demultiplication in nonlinear electrical circuits.

Nonlinearity in radio circuits makes possible, among other things, the transmission of information by amplitude modulation—what is now called AM radio (see, for instance, Marion 1970, 173–74). It also allows one to obtain, from a given oscillation, subharmonic frequencies, which are smaller than the frequency of the original wave. From a wave with frequency w, for instance, one can obtain a wave that contains the frequency $w/3$—a phenomenon Van der Pol labeled "frequency demultiplication."

In the 1927 paper reporting this phenomenon, a diagram showing the behavior of the the Van der Pol circuit appears

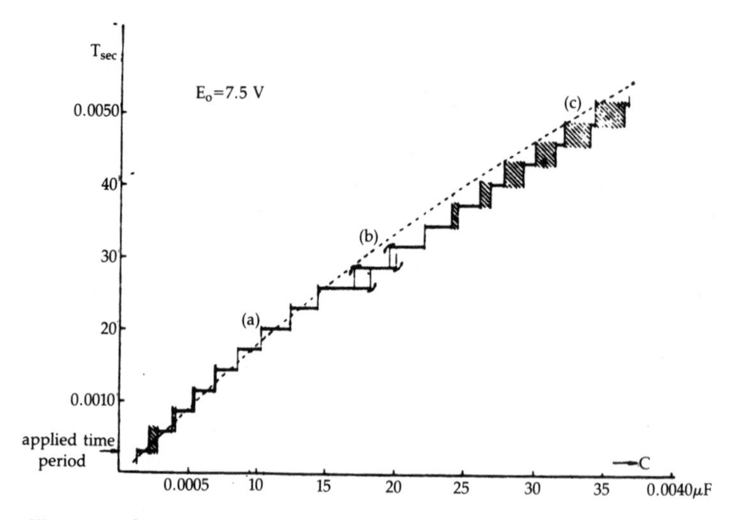

Fig. 12. Chaos in the Van der Pol circuit. Reprinted by permission from *Nature* 120:364. Copyright © 1927 Macmillan Magazines, Ltd.

(see fig. 12). Here the period T (representing the reciprocal of the frequency obtained) is plotted against the value C of a capacitor in the circuit. As C is increased, larger periods and hence smaller frequencies are found. These new frequencies sometimes occur in discrete jumps as the capacitance reaches a certain value, as we see at the point marked *a*. At *b* we note hysteresis effects, of the type discussed in chapter 4. The hysteresis receives no mention in the article.

At certain values of the capacitance we find a shaded part of the diagram, for instance at *c*. These values "correspond to those settings of the condenser where an irregular noise is heard" (p. 364). The authors remark that "often an irregular noise is heard . . . before the frequency jumps to the next lower value. However, this is a subsidiary phenomenon, the main effect being the regular frequency demultiplication" (p. 364). This "irregular noise" is chaos. Here, in a very simple electrical circuit (see fig. 13), chaotic behavior was observed, recorded, and summarily dismissed. For fifty years, this "subsidiary phenomenon" was ignored while scientists and engineers concentrated on the "main effect": the regular, predictable, and useful production of subharmonics.

Chaos could have been investigated earlier than it was. The mathematical models, the analytical tools, and the experimental devices to study nonlinear behavior were all present and available in 1920. But chaos theory did not emerge until after 1970. Why? This puzzling historical situation I have labeled the nontreatment of chaos, and scientists have puzzled over it in various ways. For example, on the deferral of experimental investigations of chaos, Bergé, Pomeau, and Vidal note that fluid dynamics does not require machinery as complex or expensive as atomic physics or cosmology; yet unlike these other fields the study of turbulence was stalled for the first sixty years of the century. "Despite its banality," they write, "this observation raises a question which historians of science will one day have to address: that of the underlying causes (circumstantial and epistemological) of the relative stagnation, in a discipline which has never lacked for practical and economic motivation" (Bergé, Pomeau, and Vidal 1984, xiii). In

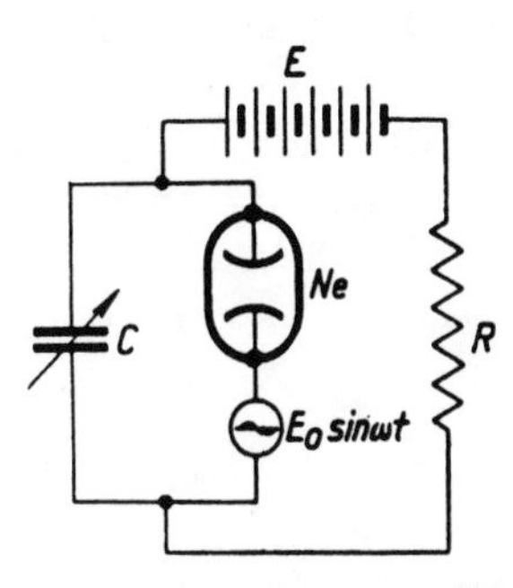

Fig. 13. The Van der Pol circuit. *C* is the variable capacitor (with capacitance measured in microfarads), *R* a resistor, and *E* a battery. $E_0\sin(\omega t)$ represents the frequency of the driving electromotive force and *Ne* is a neon glow lamp, the source of the nonlinearity. Reprinted by permission from *Nature* 120:364. Copyright © 1927 Macmillan Magazines, Ltd.

what follows, I take up these issues of the historical circumstances and epistemological predilections responsible for the nontreatment of chaos.

Waiting for the Digital Computer

Many researchers in chaos theory are aware of some of the theoretical and experimental precursors that had to wait to be developed. When asked the reason for this situation, the standard answer is that nonlinear systems do not yield straightforward mathematical solutions, so it took the digital computer to make chaos accessible to scientific inquiry. Francis Moon gives a typical formulation of this response when he writes that "specific manifestations of chaotic solutions had to wait for the arrival of powerful computers with which to calculate the long time histories necessary to observe and measure chaotic behavior" (Moon 1987, 10).

Note the strong modality of this claim: chaos "had to wait" for digital computers. John Franks makes a similarly strong claim when he proposes that "the reason why much more is known today concerning this system, the so-called Lorenz attractor, than would have been possible to know fifty years ago is, of course, the existence of digital computers" (Franks 1984, 136). Although not all writers express themselves so strongly,[3] many would agree with what I will call the computational explanation for the nontreatment of chaos: the notion that this situation is accounted for by the lack of digital computing equipment.

The Computational Explanation

John Franks articulates the most sophisticated version of the computational explanation in his 1989 review of James

3. The importance of the development of digital computers for explaining the delay of chaos is discussed in less forceful terms by Reichl and Zheng (1987, 18) and Wrightman (1985, 23).

Gleick's popular work *Chaos*. There he argues that the supposed "revolution" in science that chaos theory allegedly represents is actually "only a corollary" of the computer revolution. Chaos theory has shown us the surprising fact that sensitive dependence on initial conditions may well be very common in nonlinear systems, but this fact "surprises us because it was invisible before the computer, but with computers it is easy to see, even hard to avoid" (p. 66).

In this account, it is only with the advent of digital computers that we are even able to see chaos. The nontreatment of chaos thus should seem no more puzzling than the nontreatment of bacteria until the invention of the microscope. Until Leeuwenhoek, people saw evidence of bacterial behavior and may even have speculated about very tiny organisms, but until there were microscopes there was no possibility of a scientific examination of bacteria. Franks makes this analogy explicit:

> The computer is a viewing instrument for mathematical models that will, in the long run, be more significant than the microscope to a biologist or the telescope to an astronomer. . . . It is no more surprising that numerous types of complex dynamical phenomena have been discovered in the last twenty years than would be the discovery of numerous kinds of bacteria if thousands of biologists were, for the first time in history, given microscopes. (P. 65)

It would be foolish to suggest that digital computers have not been crucial to the development of chaos theory in its present form. Certainly the role of computational resources must be included in any acceptable explanation for the nontreatment of chaos. But this does not mean that the computational explanation is adequate on its own; such a claim makes a faulty analogy with the sorry state of the study of Pluto before the telescope. Below I will argue that the computational explanation is insufficient because of the availability of alternative computational resources.

Other Resources

I have consistently spoken of the computational explanation in terms of digital computers because not all computers are digital. Analog computers predate the digital models and this fact is the first step in refuting the sufficiency of the computational explanation. For, while it is true that nonlinear systems demand great computational resources, those resources were available well before the advent of digital computing.

Analog computing "solves" a difficult equation by constructing an electrical circuit whose elements correspond to the mathematical features of the equation.[4] From one point of view, the electrical circuit investigated by Van der Pol is precisely an analog computer for the investigation of the oscillator equation that bears his name.

Robert Shaw used an analog computer to study the behavior of the Lorenz system, producing a rough picture of the attractor and making quantitative measurements for the Lyapunov exponent of the system (Shaw 1981a, 97). He also used an analog device to study the chaotic attractor discovered by Birkhoff, producing visual displays that showed the characteristic stretching and folding of the strange attractor (Abraham and Shaw 1984). The claim that study of the chaotic systems had to wait for digital computers must therefore be modified: investigation of chaotic behavior had to wait for appropriate computational resources. These resources were available well before the digital computer arrived, in the form of analog devices.[5] In fact, Shaw contends that general purpose digital computers are not as good as analog devices for investigating wide ranges of parameters in search of an understanding of qualitative features; analog machines provide instant visual

4. A substantial discussion of the use of analog computing for the study of nonlinear systems appears in Downey and Smith (1960, 314–45).

5. While analog computing was invented at mid-century, the oscillators of Duffing and Van der Pol can be studied with simpler analog devices. While analog devices can be used for simulation, they are not very useful for data acquisition and processing.

feedback, while digital machines do not and are "horrendously expensive" besides (Shaw 1984, 19).

Even before analog computing was fully developed, Hayashi was using electrical circuits to investigate the behavior of nonlinear systems. His investigations, conducted between 1944 and 1948, revealed a striking diagram (see fig. 14). Note that while this is not in fact an instance of chaotic behavior, it does bear strong resemblance to the phenomenon of fractal basin boundaries. This phenomenon produces sensitive dependence on initial conditions, although the long-term behavior of the system is stable and periodic (see, for instance, Grebogi, Ott, and Yorke, 1987).

The availability of analog computation undercuts the sufficiency of the computational explanation, making the nontreatment of chaos seem even more puzzling. The question now becomes: given the presence of mathematical and experimental precursors to chaos theory and the possibility of studying chaos with analog computers, why did this field of research remain unexplored for so long? This question becomes even more puzzling when we consider the even earlier availability of graphical analysis, which requires no computational resources beyond the compass and straightedge.

Graphical analysis can furnish a depiction of the onset of chaos in the logistic map (see, for instance, May 1976). By constructing a diagonal line corresponding to $x = y$, and a parabola intersecting the x-axis at 0 and 1, and with height $\alpha/4$, one can iterate the logistic map by moving from the x-axis up to the diagonal, then over to the parabola, then back to the diagonal and so on. This method of the parabola and the diagonal can also be used for serious analysis.[6] The parabola, a simple polynomial curve, could be constructed by the ancient Greeks. By careful iteration, a researcher could discover the period-doubling bifurcations and the onset of highly irregular

6. Robert Shaw (1981a, 102) describes just such a possibility. The method of graphical analysis is referred to as "the diagram of Koenigs and Lemeré" in Butenin 1965, 70. For an excellent introduction to the method of graphical analysis, see Hofstadter 1981.

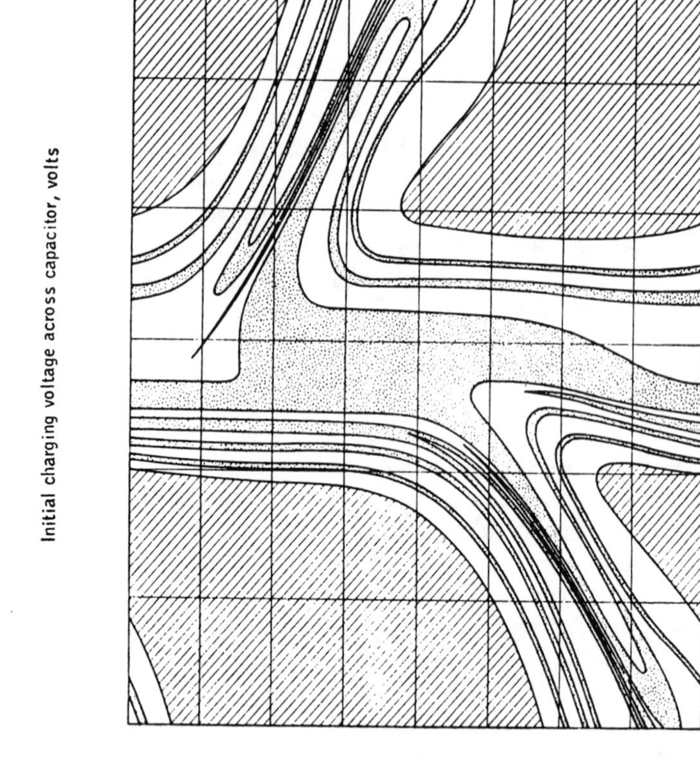

Fig. 14. An early nonlinear system. The shading indicates the regions of initial conditions which give rise to the different kinds of periodic oscillations. From Hayashi 1953

aperiodic behavior in a simple system. This fact serves to undercut the idea that the digital computer stood as a necessary condition for the development of chaos theory. The method of graphical analysis surely would not suffice to study the Lorenz system of differential equations, for instance, but it could have provided evidence for the crucial insight that simple systems can have very complicated behavior—an insight that Robert May found was still unappreciated in 1976.

Implications

Before proceeding, let us consider the way in which computational resources have influenced the historical development of science in this case. Chaos theory doubtless provides an instance of technological progress (or its lack) having a considerable effect on the direction of scientific endeavor. Without computers, research concentrated on systems that could yield exact solutions, whether by straightforward or occasionally ingenious analytic techniques. Access to digital computation allowed the widespread investigation of the nonlinear systems that manifest chaos. Indeed, as Franks suggests, once one is provided with a computer and sent into the wilds of unexplored nonlinearity, it is difficult not to find chaotic behavior wherever one turns.

Chaos theory supplies an example of how "science is often led by technical developments that open new opportunities for experimentation and theorizing rather than just by new (approximately true) theoretical insights themselves" (Rouse 1987, 140). In this case, digital computers enabled researchers to perform the rapid computations needed to investigate the solutions of nonlinear equations that do not admit of simple closed-form integrals. Without this computing power, mathematicians can study the behavior of these solutions, but it is exceedingly difficult to draw pictures of them. Researchers in chaos theory rely heavily on computer simulations of dynamical systems to supply these pictures and the new "intuition" they provide (Jensen 1987, 172).

One need not hold that chaos was impossible to study without digital computers to agree that they played the role of a crucial catalyst that impelled research in a certain direction. This catalytic effect overcame the longstanding factors that had led to the deferral of the mathematical precursors of chaos theory. Poincaré and Birkhoff knew that simple nonlinear systems can behave in complex and unpredictable ways, but this powerful idea met with neglect and, later, with resistance. Franks writes that this resistance is "surprising and fascinating" and that it was overcome only when the availability of computers made chaos "impossible to ignore" (Franks 1989, 66). But the computational explanation does not tell us what factors were responsible for this resistance. If they had been absent, could chaos theory have escaped its half-century postponement? This question verges on inadmissible counterfactual speculation, but it does point us toward an investigation of what those fascinating factors may have been.

Institutionalized Neglect of the Nonlinear

Several writers have suggested an institutional account of the factors responsible for the resistance to chaos, factors that were finally overcome when digital computers arrived. By an institutional account I mean an explanation in terms of the professional training of scientists, training that diverted attention away from areas where chaos might have been discovered and studied. By training students to disregard apparently disordered behavior, to focus attention on linear models, and to seek simple exact solutions, scientific education screened off chaos. Chaotic behavior, as I have argued, was not naturally hard to see, requiring powerful new devices to render it a possible subject of inquiry. Rather, "physicists had learned not to see chaos."[7] Education in the natural sciences created the impression that linear and solvable systems were the only ones (or at least the only important ones)—an impression that

7. This idea is attributed to James Yorke in Gleick 1987, 67.

came very close to being a prejudice in favor of systems as regular and predictable as clockwork.

Training and Invisibility

Professional instruction rendered chaos less visible in two ways: on the one hand, students were indirectly steered away from nonlinear systems (the only systems in which chaos is possible) by being taught that they were uninteresting or exceptional cases. On the other hand, when apparently bizarre or even chaotic behavior was found in these nonlinear systems, it was dismissed as mere noise or experimental error, or was "buried in a discussion of more classical nonlinear vibrations." (Moon 1987, 9). Before taking up the first of these points, I will present an argument for the latter.

To find historical examples of chaotic behavior being found and dismissed presents some difficulty. But we can find anecdotal evidence of a general kind in the scientific literature.[8] For instance, Francis Moon writes:

> Early scholars in the fields of electrical and mechanical vibrations rarely mention nonperiodic, sustained oscillations with the exception of problems relating to fluid turbulence. Yet chaotic motions have always existed. Experimentalists, however, were not trained to recognize them. Inspired by theoreticians, the engineer or scientist was taught to look for resonances and periodic vibrations in physical experiments and to label all other motions as "noise." (1987, 91)

In a similar vein, Thompson and Stewart caution researchers who may stumble across signs of a strange attractor that "such apparently random non-periodic outbursts may be the

8. Philip Marcus gives such evidence in Gleick (1987, 56) where he reports that during graduate training in physics he sometimes would stumble across behavior which he later realized showed signs of chaos. When he asked about it at the time he would be told, "Oh, it's experimental error, don't worry about it."

correct answer, and should not be attributed to bad technique and assigned to the wastepaper basket, as has undoubtedly happened in the past" (1986, xi).

Before 1970 (and even after), chaotic behavior found in a mathematical model or an experimental system might well have been dismissed as a result of professional training. An example of this training is found in the classic lectures of Hadamard from 1923, which present the problem-solving methods of classical physics in crystalline form. Here Hadamard states that the problem is incorrectly set if the solutions do not depend continuously on the initial conditions. John Earman points out that Hadamard's instruction amounts to a "methodological injunction" that a mathematical model exhibiting sensitive dependence on initial conditions must be a mistake (Earman 1986, 154; see chap. 2).

Chaos is as common as daffodils in spring (Ford 1986, 3), yet even when looking right at it, scientists often saw nothing of interest. On some occasions, students may well have been directly instructed not to pay attention to anomalous disorderly behavior. In other cases, the training had an indirect effect through the methods, concepts, and resources designated as appropriate. "You don't see something until you have the right metaphor to let you perceive it" (Robert Shaw, quoted in Gleick 1987, 262). The highly counterintuitive nature of strange attractors, which combine rapid divergence and global attraction, is regarded as a crucial factor in the nontreatment of chaos by Bergé, Pomeau, and Vidal (1984, 119). But why did scientific intuition take the shape it did? Why were the right metaphors lacking? These questions demand a further investigation of the education of natural scientists.

A Linear Prejudice

One of the most important sources of the skewed intuition and limited metaphors produced by scientific training was the overwhelming emphasis on the study of linear systems. While not all nonlinear systems display chaos, chaotic behavior can appear only in a nonlinear case, so the exclusion of nonlinearity rendered chaos much harder to identify and appreciate. In his 1976 article, considered by many a turning point in the development of chaos theory, Robert May issued an "evangelical plea" for attention to nonlinear systems:

> The elegant body of mathematical theory pertaining to linear systems (Fourier analysis, orthogonal functions, and so on), and its successful application to many fundamentally linear problems in the physical sciences, tends to dominate even moderately advanced University courses in mathematics and theoretical physics. The mathematical intuition so developed ill equips the student to confront the bizarre behaviour exhibited by the simplest of discrete nonlinear systems, such as [the logistic map]. (P. 467)

May goes on to urge that students be exposed to the logistic equation early in their education, since it can be studied with a simple hand calculator or even with pencil and paper. He contends that it would be advantageous for all of us to gain an appreciation for the role of complex nonlinear behavior in scientific as well as political and economic contexts.

While May addressed the dominance of linear models in the biological and social sciences, the situation in physics, engineering, and applied mathematics was no different. A survey of advanced mechanics textbooks published between 1910 and 1970 reveals that nonlinear systems were overwhelmingly ignored until roughly 1960 (see table). Only two of the 19 textbooks published before 1949 treated nonlinear oscillations; one of these stated that nonlinear behavior occurs "only occasionally" (Pohl 1932, 267) and the other called it an "important" topic but said "we shall not go into it in detail" (Slater and Frank 1947, 40).

We should take seriously the notion that the textbook is an important source of information about scientific theories as

Date of Publication	Nonlinearity treated	Nonlinearity not treated
1910–19	0	4
1920–29	0	4
1930–39	1	4
1940–49	1	5
1950–59	3	5
1960–69	10	8

Note: This survey includes 45 textbooks in mechanics or dynamics at the beginning graduate or advanced undergraduate level found in the Northwestern University Science and Engineering Library in 1990 and the Indiana University Science and Mathematics Library in 1991. This table indicates whether the textbook mentions nonlinearity or nonlinear oscillations in its index or treats the nonlinear (or "anharmonic") oscillator in the text.

well as about scientific education. And as Giere points out, textbooks in mechanics are typically organized according to the mathematical form of the force function (1988, 66). When discussions of oscillatory behavior treat only linear force functions, and usually fail even to mention that this is a special case, a linear prejudice can result.

Physics proceeds by the articulation and application of models, and until 1960 those models were overwhelmingly linear. Nonlinearity, which is the situation in most real systems, was usually not even mentioned. So it was only natural that "when approaching dynamic experiments in the laboratory, [scientists and engineers] looked only for phenomena that fit the linear mathematical models" (Joseph Keller, quoted in Moon 1987, 10). More anecdotal evidence for the dominance of linear models may be found in Gleick's popular account of the history of chaos theory (1987, esp. pp. 52–66). In an otherwise negative review of this book, John Franks nonetheless agrees that the emphasis on linear systems "may even have reached the point, as Gleick suggests, of scientists training themselves not to see nonlinearity in nature" (Franks 1989, 65).

The prejudice in favor of linear systems took the form of a kind of tyranny of selective example. Textbooks usually treated only linear force laws in discussions of oscillations, for instance, thus assimilating all vibrational behavior to the motion of a mass on an ideal spring—what is known as simple harmonic motion. Some textbooks mentioned that this idealization was only approximate for real systems, but justified the approximation by limiting the analysis to small amplitudes.[9] The few that treated nonlinear force laws concentrated on mathematical techniques such as perturbation theory, which aim to transform the situation into simple harmonic motion or at least reduce it to an easily solvable form.

Consider the case of the double pendulum, which consists of a light rigid rod connected by a pivot to a stationary object with another rod free to pivot at its lower end. If the double pendulum is given a small push, the two rods will swing gently in small-amplitude periodic motion. At a recent lecture, James Yorke gave such a device a gentle push and remarked, "what's in textbooks is how to study *this* solution" (Yorke 1990). He then gave the double pendulum a hefty swing, which caused it to execute exquisitely complex chaotic motion, and remarked that, apparently, until twenty years ago no one ever swung that hard.

Why should they have, when textbooks as recent as 1965 said of the double pendulum, "the interesting part of this example is the case when the angles θ_1 and θ_2 are small so that we can neglect all but the terms in the first power" (Hauser 1965, 269). By neglecting all but the linear terms, one can reduce the motion to that of two coupled simple harmonic oscillators. Anyone with college-level training in physics can appreciate the simplicity and beauty of this model system, and science must often proceed by assimilating new phenomena to well-understood models. But at least in the case of oscillations

9. But consider a particle in a potential of the form $V(x) = -V_1\exp(-ax^4)$. Here the equation of motion is $m\ddot{x} + 4aV_1x^3 = 0$. So "even for very small amplitudes the motion is not simple harmonic" (Bradbury 1968, 173–74). The assertion found in many textbooks that *all* periodic motion is simple harmonic motion if the amplitude is sufficiently small is therefore not correct.

I would contend that nonlinear phenomena have been not so much assimilated as repressed.[10]

Of course this repression was never complete. In the Soviet Union, in particular, nonlinear phenomena were the subject of interest throughout the twentieth century. But even when nonlinear systems were studied, the emphasis of the investigation was so tightly focused on periodic behavior that chaos was not likely to come into the field of attention. Those who brought nonlinear studies to the notice of the English-speaking world brought an investigation of periodic phenomena—the study of the limit cycle (see chap. 1). All efforts were focused on finding limit cycles and determining their stability (Minorsky and Leimanis 1958, 113–14; Hayashi 1953, 1).

The literature on the dynamics of a particle in a nonlinear potential relied on perturbation theory to discover the conditions that would produce stable periodic (or quasiperiodic) motion. An institutionalized reliance on these methods "prevented the discovery of very unusual effects which are associated with the global properties of the solutions" (Huberman 1981, 64). Very simple equations for a system experiencing friction, periodic forcing, and a nonlinear potential can yield chaotic solutions, but they were not found. Analogous physical systems were studied, but chaotic behavior was either not seen or not deemed worthy of attention, because of the assumption, carried over from the linear case, that a periodic force must produce periodic behavior (Moon 1987, 13; Lorenz 1963, 131).

According to Robert May, the situation was similar in the case of nonlinear difference equations in such fields as population biology and economics. "Studies of the dynamical properties of such models usually consist of finding constant equilibrium solutions, and then conducting a linearised analy-

10. Further historical research is called for to discern what direction classical physics took between 1920 and 1970. To what extent did the neglect of the nonlinear in textbooks carry over into the research projects of those physicists working on phenomena neither subatomic nor cosmological in scale?

sis to determine their stability with respect to small disturbances: explicitly nonlinear dynamical features are usually not considered" (1976, 459). In this way, science training made chaos very hard to see.

Solvability and the Clockwork Hegemony

The Art of the Integrable

The Nobel prize–winning scientist Sir Peter Medawar has called science "the art of the soluble." While this description is accurate, it does nothing to specify what counts as a solution. There was a time, for instance, when a solution to a physics problem that contained probabilistic terms was not considered an adequate solution at all, but this view has changed. In classical physics a solution meant a closed-form solution to the equations of motion: a straightforward equation which tells you the state $X(t)$ of a system at arbitrary time t, given the state $X(0)$ of the system at some initial time. Such a solution is typically a relatively simple functional relation, such that prediction and retrodiction is only a matter of "plugging in" the initial conditions and performing some more or less routine mathematical operations. If we can do this, we consider the problem solved. The existence-uniqueness property of differential equations guarantees that a system will follow exactly one trajectory in its evolution, but such an assurance gives no clue as to how to find this trajectory in the general case. So we may know that there are solutions but be unable to find the general, closed-form, exact solution.

Nonetheless, the notion arose in the eighteenth century that "all sufficiently smooth Newtonian systems were exactly and meaningfully solvable. In consequence, one no longer spoke of unsolvable systems, only those not yet solved" (Ford 1986, 24). Physicists enjoyed a great deal of success in finding such exact solutions for certain simple systems, and this led to the hope that all mechanical systems could be exactly solved (Crutchfield et al. 1986, 49). The fulfillment of this hope came in the form of an entire class of systems that were *integrable*—the differential equations have as many integrals of motion as the system has degrees of freedom (Thompson and Stewart 1986, 325). Each integral acts as a constraint on the evolution of the system, so integrable systems usually yield a closed-form solution. The stability of these solutions meant that more complicated systems could be analyzed with the approximative techniques of perturbation theory.

The earlier discussion of physicists' exclusive focus on linear systems must therefore be amended to acknowledge the attention paid by classical physics to nonlinear but integrable systems. The foremost example of such a system is the problem of two bodies under the influence of a force such as gravity, which varies as the inverse of the square of the distance between them. Such a force is not linear, but it yields an exact closed-form solution.[11] The integrable systems, yielding exact solutions or stable approximations, became "the source of almost all our textbook knowledge of classical mechanics" (Ford 1986, 25).

Prigogine and Stengers write that "integrable systems have been the model par excellence of dynamic systems, and physicists have attempted to extend the properties of what is actually a very special class of Hamiltonian equations to cover all natural processes" (1984, 74). The dominance of this class is understandable, they claim, because it is "the only one that, until recently, had been thoroughly explored."[12] Certain systems, the integrable systems of inverse-square forces without

11. In fact, the typical solution for such a system is an elliptical orbit, which is mathematically equivalent to the motion of two simple harmonic oscillators coupled together perpendicularly. Talk of a prejudice in favor of linear systems, as opposed to strictly integrable systems, is not so inaccurate after all.

12. Norman Packard supports the relevance of such an explanation when he claims, "the phenomenon of chaos could have been discovered long, long ago. It wasn't, in part because this huge body of work on the dynamics of regular motion didn't lead in that direction. But if you just look, there it is" (quoted in Gleick 1987, 251).

damping and the linear systems of harmonic motion (with and without linear damping), were considered easy to solve. Physics concentrated on these cases, treating all others as exceptions to be approximated by a simple solution or else ignored.

Yet most physical systems involve more than two bodies or nonlinear forces, so the exactly solvable cases actually comprise a tiny minority. Physicists justified their concentration on solvable systems by describing these systems as the most important ones. Consider that the force experienced by a particle might be expressed generally as $F(r) \propto r^n$, that is, depending on some power of the distance from a point of equilibrium. Only if $n = 1, -2$, or -3 can a solution be expressed in terms of the relatively simple circular functions—in all other cases a closed-form solution will have to rely on the unwieldy elliptic integrals or else it will be completely unavailable (Marion 1970, 248).

The case where $n = 1$ is just the linear oscillator, and the case $n = -2$ is the inverse-square force law of gravity. Marion goes on to claim that "these two cases, $n = 1, -2$, are the ones of prime importance in physical situations" (p. 248). Why is the case $n = -3$ not important? Perhaps because orbits under such a potential cannot yield periodic behavior (p. 251). Physics concentrates on systems that allow exact, stable, and periodic solutions because they are the most important cases, but there is a curious self-reinforcement at work here. We concentrate on exact periodic solutions because they are the most important, but might it not be the case that they are the most important in part because we have concentrated on them so exclusively?

The mutual interdependence of solvability and importance finds striking expression in Symon's (1953) mechanics textbook. There we read that "the most important problem in one-dimensional motion, and fortunately one of the easiest to solve, is the harmonic, or linear oscillator" (p. 37). Two pages later we read that "linear equations are important because there are simple general methods for solving them." The question Which comes first, solvability or importance? cannot be answered because each seems prior to the other: is the linear oscillator intrinsically important and therefore fortunately solvable, or is it easily solvable and therefore important? In what follows, I will suggest that the latter is actually the case. Judgments about linear systems parallel the situation Joseph Rouse describes with respect to early modern chemistry: "these particular things and the processes they underwent were important because they could be minutely and precisely known, rather than being known because they were important" (Rouse 1987, 229).

Reflect on the following: in conservative, or nondissipative systems, the guiding exemplar of classical mechanics was the motion of celestial bodies. Even when the three-body problem was shown not to admit any exact general solution, physicists concentrated on using perturbative methods for finding stable periodic motions.[13] In dissipative systems, the standard example was periodic behavior in the damped harmonic oscillator. Nonlinear systems, when they were acknowledged to exist, were again studied with perturbative methods in order to determine where stable periodic behavior could be found.

Stable periodic behavior is clocklike behavior. The simple harmonic oscillator keeps the same regular rhythm no matter how it is initially "wound up." A stable closed orbit in celestial mechanics, an ellipse, is equivalent to conjoined harmonic oscillators. A mechanical clock, which of necessity must be a damped, driven nonlinear oscillator, is designed to display limit cycle behavior that mimics the motion of the idealized linear case.

Physics made possible the construction of highly accurate timekeeping devices. But the appeal of stable periodic motion

13. It should be noted that even Poincaré, who first demonstrated that the three-body problem for an inverse-square force was not exactly solvable, spent much time finding stable periodic orbits. In dissipative systems, periodic motion was considered the general case to such an extent that "aperiodic" motion meant only that there was so much friction that the system immediately ground to a halt.

was somehow so great that physicists began to see everything as a clock, to the extent that nonperiodic behavior was denied or dismissed. This development I label the clockwork hegemony. Consider how much of our lives is influenced by the clocks in every room and on every wrist, or by the (stable, periodic) electromagnetic waves that carry radio and television signals, or by the motors, engines, and dynamos that rely on carefully maintained periodic motion. If physics considered that everything important is an instance of stable periodic motion, then technology has made the world we live in comply—it has filled the world with clocks.

Solvability Reconsidered

What made exact closed-form solutions and stable periodic behavior so desirable as to justify the neglect of a wide range of mathematical models and physical phenomena? For Joseph Keller, it was the "completeness and beauty" of the solutions of linear equations that led to their "domination of the mathematical training of most scientists and engineers" (quoted in Moon 1987, 91). Prigogine and Stengers speak of the "fascination" exerted by simple and exact solutions, while von Neumann cautioned that with nonlinear equations, "bad mathematical difficulties must be expected" (Prigogine and Stengers 1984, 74; Gleick 1987, 24). I contend that the intriguing but unsatisfying rhetorical opposition between fascinating beauty and bad difficulty will not suffice to account for the appeal of simple solutions.

Instead I propose a further investigation of the motivations for linear prejudice and the single-minded quest for exact solvability.[14] Such an investigation would articulate connections between certain notions of predictability, stability, and solvability—connections that explain how these conceptual

14. Speaking of a linear "prejudice" may create the impression that I believe scientists to have been unreasonable or even morally suspect, but this is not my intent. Scientific attention must be focused and restricted, but this fact should not keep us from asking why particular limitations took hold.

commitments could take hold in pedagogical and experimental practice. For the nontreatment of chaos cannot be fully accounted for by reference to the arduous and ugly mathematics required. Nonlinear systems were not neglected solely because they were so hard to study; they were hard to study in part because they had been neglected.

Before delving into the motivations for the search for simple solutions, we must acknowledge an intriguing assertion that appears in the work of some scientists who study nonlinear systems. They maintain that our very notion of simplicity is skewed unfairly in favor of linear systems and that nonlinear systems, counter to everything taught in standard textbooks, are actually simpler than linear ones.

Consider, for instance, the equation for the simple harmonic oscillator, that exemplar of linear systems: $m\ddot{x} + kx = 0$. When called upon to solve this equation, most textbooks rely on the time-honored method of just writing down the answer: $A\cos(wt + \phi)$. But where did this solution come from? And, more important, what exactly is this thing called "cosine"? A typical response makes reference to a table of trigonometric functions or an infinite series expansion or an odd-looking expression containing a transcendental number, *e,* and an imaginary number, *i.* There is no straightforward sense in which any of these explications of the cosine function is "simple," except perhaps that such functions as sine and cosine are well understood.[15] Yet much of the reason that they are so well understood may be that they have been used for so long to study the behavior of stable periodic motion.

By way of contrast, the Van der Pol oscillator yields no easy closed-form solutions but instead produces structurally stable limit cycles (and strange attractors).

> Of course, traditionally we have been led to believe that linear equations are "simpler" than non-linear equa-

15. Further discussion of the "difficulty" of solving linear systems is given in Jackson 1989, 8–11.

> tions, but this is a purely pragmatic point of view arising from a habit of giving names to the solutions of certain familiar equations (sin, cos, log) and thereby deluding ourselves into thinking that we have "solved" them. The Van der Pol equation is simpler for a much deeper mathematical reason: it preserves its quality under small perturbations. (Zeeman 1968, 5)

The relativity of judgments of simplicity becomes manifest here as we see structural stability, which allows for precise qualitative predictions, given priority over exact solvability, which allows for precise quantitative predictions.

Exact closed-form solutions, which took the shape of routine mathematical expressions, received the label "simple." Systems that required a nonstandard formula, or admitted no such formula at all, were labeled "intractable" or "impossible to solve." This labeling reinforced a linear prejudice and the attendant repression of certain phenomena. As Joseph Ford describes it, such systems were "abandoned" for failing the test of integrability; they were "exiled" for exhibiting behavior considered too complex (Ford 1987, 8). Our question now becomes: why were "simple," exact solutions the overriding aim of physics, even when the mathematical and experimental tools were available to study phenomena that did not yield this kind of solution?

Previous chapters have suggested elements of an answer to this question. The metaphysical comforts of determinism begin to explain why apparently random or "noisy" experimental results might be so easily dismissed as unsuitable for scientific investigation. The desire for exact quantitative predictability can help to account for the benefits ascribed to finding simple closed-form solutions. The notion that science should seek only straightforward causal mechanisms expressed in microreductionist, ahistorical language would obviously hinder the development of chaos theory.

All of these methodological predilections and attendant metaphysical presuppositions play a role in explaining the nontreatment of chaos. But these supplements to the computational and institutional explanations still do not yield a sufficient explanation: broader cultural or "external" factors must also play a part. In what follows, I will show that in the case of the development of chaos theory, cosmological and epistemological judgments were inextricably connected with judgments about practical utility and social interests.

A Role for Social Interests

In *Order out of Chaos,* Prigogine and Stengers discuss chemical reactions that can undergo a bifurcation and begin complex oscillatory behavior. Recent results from the investigation of these systems "could have been discovered many years ago" but were not. Although the systems themselves were found in the nineteenth century, their study was "repressed in the cultural and ideological context of those times" (1984, 19).

The repression of this phenomenon, and the recent blossoming of interest in it, resemble aspects of the historical development of chaos theory. In seeking to understand such an episode in the history of science, I am following the suggestion that "we have to incorporate the complex relations between 'internal' and 'external' determination of the production of scientific concepts" (1984, 309).[16] In what follows I will make use of some recent work in feminist philosophy of science to develop a model for the interaction between the scientific and extrascientific parts of society. This model is useful because it does not mandate a sharp division between "good science" and "ideology" but seeks to trace the workings of social and cultural influences inside scientifically respectable practices.

16. The quotation marks around the words "internal" and "external" reflect a conviction, which I share, that this dichotomy is misleading. The division between internal and external factors rests on the notion that we can distinguish purely intellectual criteria for problem selection and theory choice from practical interests. I will not argue against this notion here, however. See Harding 1986, chap. 8, and Rouse 1987, chap. 1, for a fuller discussion of the collapse of this dichotomy.

A Model from Feminist Philosophy of Science

Recent work by such authors as Evelyn Fox Keller, Sandra Harding, and Helen Longino suggests several ways to articulate further and amplify our understanding of the nontreatment of chaos. Feminist philosophy of science is one of several approaches to analyzing the role of social interests and ideology in science. While the efforts of these and other feminist writers have been focused on the biological and social sciences, their analysis can be applied to the physical sciences as well.

Feminists focus on gender ideology—unquestioned beliefs regarding the attributes and roles of women that are presumed to be "natural." The institutional focus on predictable linear systems also has ideological features, because it functions as a tacit assumption in many projects and (as I shall argue) supports certain social interests, broadly conceived. Feminist accounts can provide one useful framework for understanding how scientific development can be influenced by its ideological context.[17] This framework can be divided into three parts. Ideology affects science by influencing the choice of phenomena to be considered important, the preference for certain kinds of methods to study them, and the judgment of which results are successful.

With regard to the choice of phenomena, Harding writes that "a key origin of androcentric bias can be found in the selection of problems for inquiry, and in the definition of what is problematic about them" (1986, 25). A standard example of this is the drastic disparity of effort expended in developing birth control technology for women as opposed to men. Once certain phenomena are chosen as interesting and important, the machinery of science can produce some undeniably impressive results. This process of assigning importance takes place in the physical sciences as well.

Boyle's law, for instance, is a highly reliable account of a

17. Other work on the interaction between ideology and science includes E. A. Burtt 1980 and Georges Canguilhem 1988.

particular phenomenon: the behavior of certain substances under specific conditions. If you are interested in certain attributes of gases and how they change when manipulated in specific ways, then this phenomenon is important. But there is no such thing as an intrinsically interesting problem. "Judgments about which phenomena are worth studying, which kinds of data are significant . . . depend critically on the social, linguistic, and scientific practices of those making the judgments in question" (Keller 1985, 11). Ideology is one factor that influences the selection of some phenomena as interesting and the neglect of others as subsidiary.

Ideology operates in an implicit way, indirectly affecting the choice of the scientific methods used to study those phenomena chosen as important. The modeling process, for instance, manifests the influence of ideology: "Models are abstractions of particular aspects of an assumed reality. Into the abstractive processes go human assumptions and values as to the significant features of that reality" (Merchant 1980, 234). Helen Longino has argued that the superiority of one type of model over another is never completely determined by the available facts. Model choice provides a place where contextual features of a culture (beyond the shared norms and cognitive values of scientists) can influence science (Longino 1987, 59; 1990, chap. 9).

Mathematical methods also bear the marks of their cultural context. Harding admits that "it may be hard to imagine what gender practices could have influenced the acceptance of particular concepts in mathematics," but she argues that such influence cannot be ruled out a priori since mathematics is developed to deal with a social world (1986, 51). The historical development of nonlinear mathematical methods provides one place where it may be possible to demonstrate a cultural influence.

Such a demonstration could follow the lines of Keller's account of the puzzling "hold" on researchers exerted by certain mathematical models. In the study of the onset of aggregation and differentiation in cellular slime molds, she documents the counterproductive commitment many researchers had to models that posited central governing elements.

> To the extent that such models also lend themselves more readily to the kinds of mathematics that have been developed, we need further to ask, What accounts for the kinds of mathematics that have been developed? Mathematical tractability is a crucial issue, and it is well known that, in all mathematical sciences, models that are tractable tend to prevail. But might it not be that prior commitments (ideological, if you will) influence not only the models that are felt to be satisfying but also the very analytic tools that are developed? (Keller 1985, 155)

This passage addresses precisely the issue of the neglect of the mathematical precursors of chaos theory. This issue is related to the question of why certain results obtained from the equations of motion count as "solutions" and others do not. And this question in turn leads us to the third arena within which ideology can function, the processes by which scientists evaluate the results of inquiry and judge their success or failure. The criteria for a successful solution, a good theory, or an adequate description depend on the practices of the scientific community.[18] It is in these practices that social and political commitments and ideology exercise their effects (Keller 1985, 11).

Feminist philosophy of science provides a useful tool for an explanation of the way social interests and ideology contributed to the nontreatment of chaos. In what follows, I will develop this social explanation, focusing on the interest in the domination of nature. After arguing for the importance of this explanatory factor, I will discuss the connections between

18. Of course there are philosophical questions as to whether there can be context-independent criteria for evaluation of theories. But even granting the possible existence of such criteria, until they are universally recognized a critique of ideological distortion is still possible as well. On this issue, see Harding 1986, chap. 3.

gender ideology and the ideology of nature as a legitimate target of domination.

Usefulness

One way into a discussion of chaos theory and the domination of nature is to focus on the concrete uses of the systems studied by physics. In this section I will show that nonlinear systems were neglected in part because they were seen as being of very limited usefulness. Because no one saw useful applications for nonlinear dynamics, this field was left underdeveloped. Practical engineering concerns, an "external" factor, strongly influenced scientific judgments about which phenomena were important.

Before the study of electronic circuits, physicists interested in oscillatory behavior studied mechanical systems such as vibrating springs, beams, and cords. In such systems, nonlinearity was described as a source of "distortion," which, for instance, made mechanical governors less efficient (Pohl 1932, 45, 267). When mechanical engineers encountered nonlinearity, they sought to damp down any oscillations that might develop. As Minorsky wrote, "most of the known mechanical non-linear phenomena are of a rather undesirable, parasitic nature" (1947, 5).

Nonlinear phenomena in electrical systems, on the other hand, inspired Van der Pol and others who followed, for "electrical non-linear oscillations constitute generally useful phenomena that are utilized in radio technique, electrical engineering, television, and allied fields" (Minorsky 1947, 5). Mary Cartwright, one of the first investigators of nonlinear oscillations in the English-speaking world, writes that her attention was attracted to these phenomena by a memorandum from the Department of Scientific and Industrial Research appealing to mathematicians for assistance in problems related to radio engineering (Cartwright 1952, 86). Until there was a practical use for nonlinear systems—radio technology—the linear case was treated as the only important one.

Even when nonlinear systems were investigated, however, practical interests steered scientists away from chaotic behavior, for, in an information transmission technology, one requires a steady, predictable phenomenon that can be modified in such a way as to transmit a signal.[19] In radio and television this is the "carrier wave"—a stable periodic oscillation. The need for a steady carrier wave produced a concentration of interest in limit cycles in nonlinear systems:

> radio engineers want their systems to oscillate, and to oscillate in a very orderly way, and therefore they want to know not only whether the system has a periodic solution, but whether it is stable, what its period and harmonic content are, and how these vary with the parameters of the equation, and they sometimes want the period to be determined with a very small error. (Cartwright 1952, 84)

The interests of emerging information transmission technology exerted a strong influence on the choice of problems studied in nonlinear dynamics. In the search for stable periodic oscillations that could be reliably manipulated for radio transmissions, complex aperiodic behavior would never have been sought out for study. Chaotic oscillations, after all, appear as a source of intrinsic "noise," which cannot be used to transmit detailed information by radio, because of the amplification of small errors. The desire for predictable, controllable phenomena left chaos neglected.

A Social Explanation

One endeavor of science is widely held to be the prediction and control of natural phenomena, and chaos theory, as a scientific enterprise, contributes to this project. How, then, could

19. See, for instance, Salmon 1984, for an elucidation of what it means to transmit information. On Salmon's account, there must be a way to carry a "mark" from the transmitter in such a way as to make the mark recognizably the same to the receiver.

an interest in prediction and control have contributed to the nontreatment of chaos? The answer is that not all forms of prediction and control are the same. The study of chaos yields qualitative predictions for systems where detailed quantitative predictions are impossible. Keller makes the parallel point that not all forms of power, or control, are the same (1985, 95–114). A first formulation of a social explanation for the nontreatment of chaos would claim that a social interest in the *quantitative* prediction and *dominating* control of natural phenomena contributed to the neglect of the study of chaotic behavior.

The meaning of quantitative prediction has already been discussed. By the "dominating control" or "domination" of natural phenomena I mean the manipulation and exploitation of nonhuman nature for human ends. Not all power over our environment is domination: the engineering of genetically altered animals in order to harvest them for synthetic drugs involves domination; the use of chemicals to aid the healing process need not. To a large extent, the domination of nature has been accomplished with the use of science and technology that rely on exact quantitative prediction.

In order to bend phenomena to human needs, natural processes must be reduced in complexity and simplified into predictable, lawlike behavior. The domination of nature thus involves "making the world more predictable, by reducing the chaotic complexity of natural events and processes to regular procedures that can be controlled" (Rouse 1987, 236). Once simplified, natural events can be controlled and exploited because they behave predictably according to laws that hold for all time and space. Such phenomena can be exactly reproduced at will (Merchant 1980, 229). In this way, classical physics "provides the means for systematically acting on the world, for predicting and modifying the course of natural processes, for conceiving devices that can harness and exploit the forces and material resources of nature" (Prigogine and Stengers 1984, 37). Modern science, according to Robert Shaw, "owes its success to its ability to predict natural phenomena, thus allowing man a degree of control over his surroundings" (Shaw 1981b, 218).[20]

The connection between exact prediction and the domination of natural processes forms an important part of an explanation of the nontreatment of chaos, for in order to satisfy a cultural and economic interest in dominating nature, science sought predictable, regular behavior. These interests help account for the fact that exact closed-form solutions were considered the premier type of solution to a scientific problem, which in turn helps to explain the prejudice in favor of linear systems. Chaotic behavior was screened off from study by a procrustean form of attention that was paid only to elements of the world that could be reduced to objects for human use: "modern natural science has as its main goal prediction, i.e., the power to manipulate objects in such a way that certain predicted events will happen. This means that only those aspects of the object are deemed relevant which make it suitable for such manipulation or control" (Schachtel 1959, 171, quoted in Keller 1985, 120).

This passage overstates the link between exact quantitative prediction and the domination of nature; it completes a brief sketch of the way some philosophers and scientists have portrayed this connection. Yet the social explanation outlined above should supplement the computational and institutional explanations. The notion that a solution to a physics problem must be a simple exact formula took hold, in part, because of a specific social interest.

Ideology and Metaphor

The feminist account of gender ideology is connected to a social explanation for the nontreatment of chaos because the ideology of nature as a legitimate target of domination shares

20. I have retained the noninclusive language used by Shaw.

something in common with gender ideology. The connection comes from a consideration of metaphorical links between the domination of nature and the domination of women.

The images conjured up by words such as "turbulence" and "chaos" wield so much power that the symbolic and metaphorical dimension of the development of nonlinear dynamics should not be overlooked. Carolyn Merchant, a historian of science, has pointed out that at the time of the rise of modern science women became strongly linked to the wild, disorderly features of the natural world perceived as needing to be subdued (Merchant 1980, 132). Even more telling perhaps is Machiavelli's comparison of fortune to a violent, turbulent river and his injunction, "fortune is a woman and it is necessary if you wish to master her, to conquer her by force" (quoted in Merchant 1980, 130). These metaphorical links pose the possibility that the denigration of women as fickle and the scientific fixation on stable predictable phenomena are intertwined.

One of the strongest metaphors operating in science is the image of the universe as a machine and the scientist as an investigator seeking to discover its hidden workings. This picture of the world arose during the scientific revolution and answered pressing needs for social order, and it bequeathed to science a concern for certainty, law, and predictability (Merchant 1980, 215, 227; see also Toulmin 1989, 45–89). But the vision of the world as a clockwork mechanism implies that physical matter is inert and dead. Thus the metaphor of the world as machine "functioned as a justification for power and dominion over nature" (Merchant 1980, 215).

The mechanistic view of the world served as a legitimating ideology for the project of dominating nature, while at the same time functioning to secure a hierarchical social order. In the context of the social upheaval of the seventeenth century, the vitalist idea that matter is alive "could be seen as aiding and abetting disorder and chaos" (Merchant 1980, 195). Elizabeth Potter argues for the influence of this opposition on such early modern scientists as Robert Boyle. If the particles of matter were endowed with life by the Creator, then even the lowliest bits of the world merit some respect. This doctrine had powerful implications for those working for a less hierarchical social order (Potter 1988, 19–34).

So the metaphor of the world as mechanism, a presupposition that colors much of the pursuit of scientific knowledge, has a social meaning as well. Here is one of the ways in which "the connection between what counts as knowledge and the ability to manipulate and control the things known, is culture bound and gender bound" (Rouse 1987, 256). Ideologies legitimating the domination of nature and the domination of specific social groups are connected.

Karen Warren has characterized this connection in terms of the common reliance on a "logic of domination"—an assumption that hierarchical differences justify subordination (Warren 1990, 128–29). Men are presumed to be more rational than women, therefore women may be subordinated. Humans are purposive while natural processes are presumably not, therefore nature may be subordinated. An analysis of the domination of nature shares much in common with an analysis of the domination of women and colonized peoples. They all seek to trace the ways ideological characterizations of hierarchy function in concert with a logic of domination.

Limitations and Prospects

The fifty-year nontreatment of chaos may not seem historically significant in light of the fact that many scientific discoveries have taken even longer to be appreciated. The Copernican theory of the solar system, for instance, took much more time to be accepted. But the rapid pace of scientific progress in the twentieth century, with a large population of professionals with high-speed communications and broad publishing resources, makes the long postponement of the study of chaotic phenomena problematic. I have discussed three explanatory factors that together can provide the beginnings of an adequate account for this situation. Some mention should be made of other historical features that deserve further attention.

The Russian literature on nonlinear phenomena may yield additional early examples of the discovery of chaotic behavior. The political and linguistic barriers between research conducted in Russian and that conducted in English call for more investigation as possible factors in the nontreatment of chaos.

During the period of the twenties and thirties physics was understandably occupied with the birth of quantum theory. Some of the reason for the nontreatment of chaos may be that *all* classical phenomena were "put on hold" while physics plunged into the study of the subatomic. This emphasis on the very small has persisted, and merits examination.

Finally, the historical question raised earlier and set aside—what accounts for the current explosion of interest in chaos?—cannot be strictly separated from the question of the nontreatment of chaos. As one scientist puts it, "how was it that after ignoring the development of analytical mechanics for half a century, physicists took it up again and cultivated it to its present state in which it is dreadfully fashionable?" (Wrightman 1985, 22). In light of the factors responsible for the neglect of chaotic phenomena, one may well ask how chaos theory could ever have broken free and reached its current success? Keller supplies the beginnings of an answer when she insists that the actual practice of science is "always more abundant than its ideology" (1985, 136). A fuller historical understanding of the development of chaos theory would require consideration of both its long germination and its subsequent blossoming.

References

Abraham, R., and C. Shaw. 1982. *Dynamics: The Geometry of Behavior.* Part 1: *Periodic Behavior.* Santa Cruz: Aerial Press.

———. 1984. *Dynamics: The Geometry of Behavior.* Part 2: *Chaotic Behavior.* Santa Cruz: Aerial Press.

Abraham, R. 1985. "Is There Chaos without Noise?" In P. Fischer and W. R. Smith, eds., *Chaos, Dynamics, and Fractals,* 189–96. New York: Marcel Dekker.

Albano, A. M., et al. 1986. "Lasers and Brains: Complex Systems with Low-dimensional Attractors." In G. Mayer-Kress, ed., *Dimensions and Entropies in Chaotic Systems,* 231–40. Berlin: Springer-Verlag.

Alexander, H. G., ed. 1956. *The Leibniz-Clarke Correspondence.* New York: Harper & Row.

Arneodo, A., P. Coullet, C. Tresser, A. Libchaber, J. Maurer, and D. d'Humiers. 1983. "On the Observation of an Uncompleted Cascade in a Rayleigh-Bénard Experiment." *Physica* 6D:385–92.

Bachelard, G. 1984. *The New Scientific Spirit.* Trans. A. Goldhammer. Boston: Beacon Press.

Barnsley, M., and S. Demko, eds. 1986. *Chaotic Dynamics and Fractals.* Orlando: Academic Press.

Bartholomew, D. 1984. *God of Chance.* London: SCM Press.

Batterman, R. W. 1991. "Randomness and Probability in Dynamical Theories: On the Proposals of the Prigogine School." *Philosophy of Science* 58: 241–63.

Bell, J. S. 1987. *Speakable and Unspeakable in Quantum Mechanics.* Cambridge: Cambridge University Press.

Bergé, P., Y. Pomeau, and C. Vidal. 1984. *Order within Chaos.* Trans. L. Tuckerman. Paris: J. Wiley & Sons.

Berry, M. V. 1986. "The Unpredictable Bouncing Rotator: A Chaology Tutorial Machine." In S. Diner, D. Fargue, and G. Lochak, eds., *Dynamical Systems: A Renewal of Mechanism*, 3–12. Singapore: World Scientific.

———. 1987. "Quantum Chaology." In M. V. Berry, I. C. Percival, and N. O. Weiss, eds., *Dynamical Chaos*, 183–98. Princeton: Princeton University Press.

Bradbury, T. C. 1968. *Theoretical Mechanics*. New York: J. Wiley & Sons.

Braun, M. 1975. *Differential Equations and Their Applications*. New York: Springer-Verlag.

Briggs, J., and F. D. Peat. 1989. *Turbulent Mirror*. New York: Harper & Row.

Burtt, E. A. [1952] 1980. *The Metaphysical Foundations of Modern Science*. Atlantic Highlands, N.J.: Humanities Press.

Butenin, N. V. 1965. *Elements of the Theory of Nonlinear Oscillations*. New York: Blaisdell.

Canguilhem, G. 1988. *Ideology and Rationality in the History of the Life Sciences*. Trans. A. Goldhammer. Cambridge, Mass.: MIT Press.

Cartwright, M. L. 1952. "Nonlinear Vibrations: A Chapter in Mathematical History." *Mathematical Gazette* 35:80–88.

Causey, R. L. 1969. "Polanyi on Structure and Reduction." *Synthese* 20:230–37.

Chandra, J., ed. 1984. *Chaos in Nonlinear Dynamical Systems*. Philadelphia: Society for Industrial and Applied Mathematics.

Chirikov, B. 1979. "A Universal Instability of Many-dimensional Oscillator Systems." *Physics Reports* 52:263–379.

Cohen-Tannoudji, C., B. Diu, and F. Laloë. 1977. *Quantum Mechanics*, vol. 1. Paris: Hermann.

Collet, P., and J. Eckmann. 1980. *Iterated Maps on the Interval as Dynamical Systems*. Boston: Birkhäuser.

Conrad, J. [1914] 1984. *Chance: A Tale in Two Parts*. London: Hogarth Press.

Cortázar, J. 1966. *Hopscotch*. Trans. G. Rabassa. New York: Pantheon.

Crutchfield, J., J. D. Farmer, N. Packard, and R. Shaw. 1986. "Chaos." *Scientific American* 255 (December): 46–57.

Devaney, R. 1986. *An Introduction to Chaotic Dynamical Systems*. Menlo Park: Benjamin/Cummings Publishing Co.

Dewey, J. 1958. *Experience and Nature.* New York: Dover.
Downey, G. L., and E. M. Smith. 1960. *Advanced Dynamics for Engineers.* Scranton: International Textbook.
Dwoyer, D. L., M. Y. Hussaini, and R. G. Voight. 1985. *Theoretical Approaches to Turbulence.* New York: Springer-Verlag.
Dyke, C. 1990. "Strange Attraction, Curious Liaison: Clio Meets Chaos." *Philosophical Forum* 21:369–92.
Earman, J. 1986. *A Primer on Determinism.* Dordrecht: D. Reidel.
Ezeabasili, N. 1977. *African Science: Myth or Reality?* New York: Vantage.
Feigenbaum, M. 1978. "Quantitative Universality for a Class of Nonlinear Transformations." *Journal of Statistical Physics* 19: 25–52.
Fine, A. 1971. "Probability in Quantum Mechanics and Other Statistical Theories." In M. Bunge, ed., *Problems in the Foundations of Physics,* 79–92. New York: Springer-Verlag.
Fischer, P., and W. R. Smith. 1985. *Chaos, Dynamics, and Fractals.* New York: Marcel Dekker.
Ford, J. 1983. "How Random Is a Coin Toss?" *Physics Today* 36:40–47.
———. 1986. "Chaos: Solving the Unsolvable, Predicting the Unpredictable!" In M. Barnsley and S. Demko, eds., *Chaotic Dynamics and Fractals,* 1–52. Orlando: Academic Press.
———. 1987. "Directions in Classical Chaos." In Hao B.-L., ed., *Directions in Chaos,* 1–16. Singapore: World Scientific.
Franks, J. 1984. Review of *Chaos: Making a New Science,* by James Gleick. *Bulletin of the American Mathematical Society,* n.s., 10: 135–39.
———. 1989. Review of *Chaos: Making a New Science,* by James Gleick. *Mathematical Intelligencer* 11:65–71.
Froehling, H., J. Crutchfield, D. Farmer, N. Packard, and R. Shaw. 1981. "On Determining the Dimension of Chaotic Flows." *Physica* 3D:605–17.
Giere, R. 1988. *Explaining Science.* Chicago: University of Chicago Press.
Gigerenzer, G., Z. Swijtink, T. Porter, L. Daston, J. Beatty, and L. Kruger. 1989. *The Empire of Chance.* Cambridge: Cambridge University Press.
Glass, L., and M. C. Mackey. 1988. *From Clocks to Chaos.* Princeton: Princeton University Press.

Gleick, J. 1987. *Chaos: Making a New Science*. New York: Viking Penguin.

Glymour, C. 1971. "Determinism, Ignorance, and Quantum Mechanics." *Journal of Philosophy* 68:744–51.

Gollub, J., and H. L. Swinney. 1975. "Onset of Turbulence in a Rotating Fluid." *Physical Review Letters* 48:927–30.

Grassberger, P., and I. Procaccia. 1983. "Characterization of Strange Attractors." *Physical Review Letters* 50:346–49.

Grebogi, C., E. Ott, and J. Yorke. 1987. "Chaos, Strange Attractors, and Fractal Basin Boundaries in Nonlinear Dynamics." *Science* 238:632–38.

Gutzwiller, M. C. 1992. "Quantum Chaos." *Scientific American* 266 (January):78–84.

Hacking, I. 1975. *The Emergence of Probability*. Cambridge: Cambridge University Press.

Hadamard, J. 1952. *Lectures on Cauchy's Problem in Linear Partial Differential Equations*. New York: Dover.

Haken, H., ed. 1981. *Chaos and Order in Nature*. Berlin: Springer-Verlag.

Hao B.-L., ed. 1984. *Chaos*. Singapore: World Scientific Publishing Company.

Harding, S. 1986. *The Science Question in Feminism*. Ithaca: Cornell University Press.

———. 1987. Introduction. In S. Harding, ed., *Feminism and Methodology*, 1–15. Bloomington: Indiana University Press.

Hauser, W. 1965. *Introduction to the Principles of Mechanics*. Reading: Addison-Wesley.

Hayashi, C. 1953. *Forced Oscillations in Nonlinear Systems*. Osaka: Nippon Printing and Publishing Company.

Hayles, N. K. 1990. *Chaos Bound*. Ithaca: Cornell University Press.

Hénon, M. 1976. "A Two-Dimensional Mapping with a Strange Attractor." *Communications in Mathematical Physics* 50:69–77.

Hirsch, M. 1984. "The Dynamical Systems Approach to Differential Equations." *Bulletin of the American Mathematical Society*., n.s., 11:1–64.

———. 1985. "The Chaos of Dynamical Systems." In P. Fischer and W. R. Smith, eds., *Chaos, Fractals, and Dynamics*, 189–96. New York: Marcel Dekker.

Hobbs, J. 1991. "Chaos and Indeterminism." *Canadian Journal of Philosophy* 21:141–64.

Hofstadter, D. R. 1981. "Strange Attractors: Mathematical Patterns Delicately Poised between Order and Chaos." *Science* 245 : 5–22.

Holden, A. V., ed. 1986. *Chaos.* Princeton, N.J.: Princeton University Press.

Holloway, G., and B. West, eds. 1984. *Predictability of Fluid Motions.* New York: American Institute of Physics.

Horwich, P. 1987. *Assymetries in Time.* Cambridge, Mass.: MIT Press.

Huberman, B. 1981. "Turbulence and Scaling in Solid-state Physics." In H. Haken, ed., *Chaos and Order in Nature,* 64–68. Berlin: Springer-Verlag.

Hunt, G. M. K. 1987. "Determinism, Predictability and Chaos." *Analysis* 47 : 129–33.

Husserl, E. 1970. *The Crisis of European Sciences and Transcendental Phenomenology.* Trans. D. Carr. Evanston: Northwestern University Press.

Jackson, E. A. 1989. *Perspectives of Nonlinear Dynamics,* vol. 1. Avon: Cambridge University Press.

Jammer, M. 1974. *The Philosophy of Quantum Mechanics.* New York: J. Wiley & Sons.

Jensen, R. 1987. "Classical Chaos." *American Scientist* 75 : 168–81.

Kadanoff, L. P. 1983. "Roads to Chaos." *Physics Today,* December 1983, pp. 46–53.

———. 1985. "Applications of Scaling Ideas to Dynamics." In G. Velo and A. S. Wrightman, eds., *Regular and Chaotic Motions in Dynamic Systems,* 27–72. New York: Plenum Press.

Kant, I. [1783] 1950. *Prolegomena to Any Future Metaphysics.* Trans. L. W. Beck. Indianapolis: Bobbs-Merrill.

Keller, E. F. 1985. *Reflections on Gender and Science.* New Haven: Yale University Press.

Kellert, S. H., M. A. Stone, and A. Fine. 1990. "Models, Chaos and Goodness of Fit." *Philosophical Topics* 18 : 85–105.

Kim, Y. S., and W. W. Zachary, eds. 1987. *The Physics of Phase Space.* Berlin: Springer-Verlag.

Kitcher, P. 1989. "Explanatory Unification and the Causal Structure of the World." In P. Kitcher and W. Salmon, eds., *Scientific Explanation,* 410–505. Minneapolis: University of Minnesota Press.

Kitcher, P., and W. Salmon, eds. 1989. *Scientific Explanation.* Minneapolis: University of Minnesota Press.

Kuhn, T. 1962. *The Structure of Scientific Revolutions*. Chicago: University of Chicago Press.

Kuramoto, Y., ed. 1984. *Chaos and Statistical Methods*. Berlin: Springer-Verlag.

Landau, L. 1944. "On the Problem of Turbulence." *C. R. Academy of Science, URSS*. 44:311–16.

Lanford, O. E. 1982. "A Computer-assisted Proof of the Feigenbaum Conjectures." *Bulletin of the American Mathematical Society* 6:427–34.

Libchaber, A., S. Fauve, and C. Laroche. 1983. "Two-Parameter Study of the Routes to Chaos." *Physica* 7D:73–84.

Libchaber, A., and J. Maurer. 1982. "A Rayleigh-Bénard Experiment: Helium in a Small Box." In T. Riste, ed., *Nonlinear Phenomena at Phase Transitions and Instabilities*, 259–86. New York: Plenum.

Livi, R., S. Ruffo, S. Ciliberto, and M. Buiatti, eds. 1988. *Chaos and Complexity*. Singapore: World Scientific.

Longino, H. 1987. "Can There Be a Feminist Science?" *Hypatia* 2:51–64.

———. 1990. *Science as Social Knowledge*. Princeton: Princeton University Press.

Lorenz, E. 1963. "Deterministic Nonperiodic Flow." *Journal of the Atmospheric Sciences* 20:130–41.

Luenberger, D. G. 1979. *Introduction to Dynamical Systems*. New York: J. Wiley & Sons.

Malament, D. 1985. "'Time Travel' in the Gödel Universe." In P. D. Asquith and P. Kitcher, eds., *Proceedings of the Philosophy of Science Association 1984 Meeting* 2:91–100. East Lansing: Philosophy of Science Association.

Marchal, C. 1988. "Qualitative Analysis in the Few Body Problem." In M. J. Valtonen, ed., *The Few Body Problem*, 5–15. Dordrecht: Kluwer Academic.

Marion, J. 1970. *Classical Dynamics of Particles and Systems*. New York: Academic Press.

May, R. 1976. "Simple Mathematical Models with Very Complicated Dynamics." *Nature* 261:459–67.

———. 1987. "Chaos and the Dynamics of Biological Populations." In M. V. Berry, I. C. Percival, and N. O. Weiss, eds., *Dynamical Chaos*, 27–43. Princeton: Princeton University Press.

Mayer-Kress, G., ed. 1986. *Dimensions and Entropies in Chaotic Systems*. Berlin: Springer-Verlag.

Menasché, J. S. 1987. Fragments from a work in progress. Personal Communication.

Merchant, C. 1980. *The Death of Nature*. New York: Harper & Row.

Minorsky, N. 1947. *Introduction to Non-Linear Mechanics*. Ann Arbor: Edwards Brothers.

Minorsky, N., and E. Leimanis. 1958. *Dynamics and Nonlinear Mechanics*. New York: J. Wiley & Sons.

Mira, C. 1986. "Some Historical Aspects Concerning the Theory of Dynamic Systems." In S. Diner, D. Fargue, and G. Lochak, eds., *Dynamical Systems: A Renewal of Mechanism*, 250–61. Singapore: World Scientific.

Misra, B., I. Prigogine, and M. Courbage. 1979. "From Deterministic Dynamics to Probabilistic Description." *Physica* 98A:1–26.

Monod, J. 1971. *Chance and Necessity*. New York: Alfred A. Knopf.

Moon, F. 1987. *Chaotic Vibrations*. New York: J. Wiley & Sons.

Newhouse, S. E., D. Ruelle, and F. Takens. 1978. "Occurrence of Strange Axiom A Attractors near Quasi-Periodic Flow on *Tm* ($m=3$ or more)." *Communications in Mathematical Physics* 64:35–43.

Packard, N., J. Crutchfield, J. Farmer, and R. Shaw. 1980. "Geometry from a Time Series." *Physical Review Letters* 45:712–16.

Pohl, R. W. 1932. *Physical Principles of Mechanics and Acoustics*. London: Blackie & Son.

Popper, K. 1956. *The Open Universe*. Totowa, N.J.: Rowman & Littlefield.

———. 1972. *Objective Knowledge*. Oxford: Oxford University Press.

Potter, E. 1988. "Modeling the Gender Politics in Science." *Hypatia* 3:19–34.

Prigogine, I. 1980. *From Being to Becoming*. San Francisco: W. H. Freeman & Company.

Prigogine, I., and C. George. 1983. "The Second Law as a Selection Principle." *Proceedings of the National Academy of Sciences* 80:4590–94.

Prigogine, I., and I. Stengers. 1984. *Order out of Chaos*. New York: Bantam Books.

Railton, P. 1981. "Probability, Explanation, and Information." *Synthese* 48:233–56.
Redhead, M. 1987. *Incompleteness, Nonlocality, and Realism.* Oxford: Clarendon Press.
Reichl, L. E., and W. M. Zheng. 1987. "Nonlinear Resonance and Chaos in Conservative Systems." In B.-L. Hao, ed., *Directions in Chaos,* 17–30. Singapore: World Scientific.
Rescher, N. 1970. *Scientific Explanation.* New York: Free Press.
Robinson, C. 1989. "Homoclinic Bifurcation to a Transitive Attractor of Lorenz Type." *Nonlinearity* 2:495–518.
Roqué, A. J. 1988. "Non-linear Phenomena, Explanation, and Action." *International Philosophical Quarterly* 28:247–55.
Rouse, J. 1987. *Knowledge and Power.* Ithaca: Cornell University Press.
Roux, J.-C., R. H. Simoyi, and H. L. Swinney. 1983. "Observation of a Strange Attractor." *Physica* 8D:257–66.
Ruelle, D. 1981. "Differentiable Dynamical Systems and the Problem of Turbulence." *Bulletin of the American Mathematical Society,* n. s., 5:29–42.
Ruelle, D., and F. Takens. 1971. "On the Nature of Turbulence." *Communications in Mathematical Physics* 20:167–92.
Salmon, W. 1971. "Determinism and Indeterminism in Modern Science." In J. Feinberg, ed., *Reason and Responsibility,* 331–46. Encino: Dickenson Publishing Co.
———. 1984. *Scientific Explanation and the Causal Structure of the World.* Princeton: Princeton University Press.
———. 1989. "Four Decades of Scientific Explanation." In P. Kitcher and W. Salmon, eds., *Scientific Explantion,* 3–219. Minneapolis: University of Minnesota Press.
Saltzman, B. 1962. "Finite Amplitude Free Convection as an Initial Value Problem I." *Journal of the Atmospheric Sciences* 19: 329–41.
Shaw, R. 1981a. "Strange Attractors, Chaotic Behavior, and Information Flow." *Zeitschrift für Naturforschung* 36a:80–112.
———. 1981b. "Modeling Chaotic Systems." In H. Haken, ed., *Chaos and Order in Nature,* 218–31. New York: Springer-Verlag.
———. 1984. *The Dripping Faucet as a Model Chaotic System.* Santa Cruz: Aerial Press.
Shimada, I., and T. Nagashima. 1979. "A Numerical Approach to

Ergodic Problem of Dissipative Dynamical Systems." *Progress in Theoretical Physics* 61:1605–16.

Skarda, C. A., and W. J. Freeman. 1987. "How Brains Make Chaos in order to Make Sense of the World." *Behavioral and Brain Sciences* 10:161–95.

Slater, J., and N. Frank. 1947. *Mechanics.* New York: McGraw-Hill.

Sparrow, C. 1986. "The Lorenz Equations." In A. V. Holden, ed., *Chaos,* 111–34. Princeton: Princeton University Press.

Stone, M. 1989. "Chaos, Prediction, and Laplacean Determinism." *American Philosophical Quarterly* 26:123–31.

Suppe, F. 1977. *The Structure of Scientific Theories.* Urbana: University of Illinois Press.

Suppes, P. 1984. *Probabilistic Metaphysics.* Padstow: Basil Blackwell.

Symon, K. 1953. *Mechanics.* Cambridge, Mass.: Addison-Wesley.

Takens, F. 1981. "Detecting Strange Attractors in Turbulence." In D. A. Rand and L. S. Young, eds., *Lecture Notes in Mathematics 898,* 366–81. Berlin: Springer-Verlag.

Teller, P. 1979. "Quantum Mechanics and the Nature of Continuous Physical Magnitudes." *Journal of Philosophy* 76:345–61.

———. 1986. "Relational Holism and Quantum Mechanics." *British Journal for the Philosophy of Science* 37:71–81.

Thompson, J. M. T., and H. B. Stewart. 1986. *Nonlinear Dynamics and Chaos.* New York: J. Wiley & Sons.

Toulmin, S. 1961. *Foresight and Understanding.* New York: Harper & Row.

———. 1990. *Cosmopolis: The Hidden Agenda of Modernity.* New York: Free Press.

Valtonen, M. J., ed. 1988. *The Few Body Problem.* Dordrecht: Kluwer Academic Publishers.

Van der Pol, B., and J. Van der Mark. 1927. "Frequency Demultiplication." *Nature* 120:363–64.

van Fraassen, B. C. 1980. *The Scientific Image.* Oxford: Oxford University Press.

Velo, G., and A. S. Wrightman, eds. 1985. *Regular and Chaotic Motions in Dynamic Systems.* New York: Plenum Press.

Von Plato, J. 1982. "Probability and Determinism." *Philosophy of Science* 49:51–66.

Walker, G. H., and J. Ford. 1979. "Amplitude Instability and Ergodic Behavior for Conservative Nonlinear Oscillator Systems." *Physical Review* 188:416–32.

Warren, K. 1990. "The Power and the Promise of Ecological Feminism." *Environmental Ethics* 12:125–46.

Wisdom, J. 1988. "Some Aspects of Chaotic Behavior in the Solar System." In M. J. Valtonen, ed., *The Few Body Problem,* 417–19. Dordrecht: Kluwer Academic Publishers.

Wittgenstein, L. 1958. *Philosophical Investigations.* Trans. G. E. M. Anscombe. New York: Macmillan.

Woodward, J. 1989. "The Causal Model of Explanation." In P. Kitcher and W. C. Salmon, eds., *Scientific Explanation,* 357–83. Minneapolis: University of Minnesota Press.

Wrightman, A. S. 1985. "Introduction to the Problems." In G. Velo and A. S. Wrightman, eds., *Regular and Chaotic Motions in Dynamic Systems,* 1–26. New York: Plenum Press.

Yorke, J. 1990. "Numerically Generated Chaotic Trajectories." Talk given at Northwestern University Mathematics Department, April 18, Evanston, Illinois.

Zeeman, E. C. 1968. *Lecture Notes on Dynamical Systems.* Aarhuis Universitet Matematisk Institut: Nordic Summer School in Mathematics.

Index